Energy Management

Energy Management

Big Data in Power Load Forecasting

Valentin A. Boicea

CRC Press is an imprint of the
Taylor & Francis Group, an **informa** business

First edition published 2021
by CRC Press
6000 Broken Sound Parkway NW, Suite 300, Boca Raton, FL 33487-2742

and by CRC Press
2 Park Square, Milton Park, Abingdon, Oxon, OX14 4RN

CRC Press is an imprint of Taylor & Francis Group, LLC

Library of Congress Cataloging-in-Publication Data
A catalog record has been requested for this book

ISBN: 978-0-367-70616-6 (hbk)
ISBN: 978-0-367-70662-3 (pbk)
ISBN: 978-1-003-14721-3 (ebk)

Typeset in Minion
by MPS Limited, Dehradun

Contents

Acknowledgments

Amicis et Collegis Italicis Germanicis et Daco Romanis Grato Animo
Iucundi Acti Labores

Thanks to the Italian, German and Romanian colleagues
It is good to carry out the job (Cicero)

Author Biography

Adrian–Valentin Boicea, a former PhD student at Politecnico di Torino, Italy, received the BS degree in electrical engineering and electrical power systems from the University Politehnica of Bucharest (UPB), Romania. At present, he is a lecturer in the Department of Electrical Power Systems at the UPB. His research interests include distributed generation systems, energy efficiency, renewable sources, electrical measurements, the Operational Research algorithms used in power engineering, and the Artificial Intelligence techniques for Big Data processing applied in the energy sector.

CHAPTER 1

Introduction

Even if it is a new concept, "Industry 4.0" has come to be known worldwide. This concept denotes the ongoing industrial revolution caused by the continuous increase in the volume of information aimed at improving traditional manufacturing and industrial practices at the global level. Previously, one spoke about the First Industrial Revolution in which steam and water power played a key role as far as production was concerned. Then, the Second Industrial Revolution, with the introduction of the telegraph, opened the door to what was to become the communication sector, which facilitated an easier transmission of ideas. Simultaneously to this, the increased electrification of the factories improved the production lines. The Third Industrial Revolution occurred, practically, toward the end of the 20th century when the electronic computers appeared. This was also named the Digital Revolution, and it continues to this day and marks the connection with the "Industry 4.0". The main differences between the two lie not in scope, which is practically the same (namely the improvement of the current industrial practices), but in the data volume and processing speed. It is expected that by 2025, the data volume at the global level will reach 175 ZB (1 ZB = 10^{21} B) while the annual growth rate of the same parameter is valued at 61% [1]. It is expected that this explosion in information quantity will engender economic growth coupled with a strong position on the market for those companies which will be able to process it.

But what is in fact generating this data deluge? The answer to this question is not too complex and is, basically, represented by the Internet of Things (IoT) sensors, GSM networks, and computer networks (including the Internet). In some way or another, these are employed in various key sectors like healthcare, academia, smart manufacturing, social media, and ultimately within the Smart City.

For instance, in the healthcare field, smartphones and smartwatches are very important when it comes to delivering certain medical parameters of the person wearing them. This has to be made in compliance with the most up-to-date data security and privacy standards. Other general utilization possibility would be the prediction of pandemics.

In academia, on the other hand, Big Data can facilitate better tailoring of the course plans for the professors as well as of the curricula for a certain specialization in the university [2].

Data-driven manufacturing enables fast decision-making based on the IoT sensors installed within the production lines [3]. Thus, a more accurate prediction with respect to quantity can be obtained. A direct consequence to this prediction would be a reduction in machine maintenance time by 20–50 % according to [3] as well as of the costs by 5–10 %. At the same time, the overall process efficiency will increase by 10–20 %.

When it comes to social media, one can say that this revolutionized the transmission of ideas, after the introduction of the telegraph, telephone, and Internet. Currently, it facilitates remote work through various dedicated platforms. In this way, the industry continues to function even if physical large gatherings are prohibited. Another advantage is the ease of tracking down various decisions made previously during the process of working remotely.

Smart Cities, on the other hand, play a vital role in the improvement of the urban life standard. This encompasses many aspects such as transportation, public safety, optimized utility, and health services. All these are obtained through the collection and processing of Big Data resulting from the IoT sensors or computer networks within the city. Practically, a Smart City consists of multiple layers of sensors, edge devices (used for the connection to the provider core networks), and platforms (for running various Artificial Intelligence – AI – algorithms) [4].

In general, Smart Grids (SG) that use various digital technologies in order to optimize power transmission are very important when it comes to the cities of the future (Smart Cities). Since a Smart City plays a key role as far as economic welfare is concerned, and because Big Data, efficient energy consumption, and, implicitly, accurate load forecast are involved with this concept, a more detailed analysis will be carried out in the sequel.

From the point of view of the data flow, the Smart City can be seen as a three-tiered structure [3]: the lowest level comprises the data ingestion and communication devices (in fact, the IoT systems), the secondary level includes different datastore and processor nodes, while the third and highest layer consists of real-time components that facilitate the information sharing between the parties as well as Big Data analytics that optimizes the various actions with regard to decision making, based on historical data.

The three-use cases for the Smart City are directly related to energy and energy efficiency. These are: the efficient delivery of citizen services (including utilities), the environmental impact (assessed also as a function of Carbon emissions), and incident management [3].

Last but not least, Smart City means achieving more with efficient energy consumption. That is why the load forecast plays a key role when considering this concept.

Hence, in section 3.2 of this book, a completely new idea of obtaining a load forecast at the national power system transmission level is presented. This idea is compatible with Big Data collection and processing, being based on a Convolutional Neural Network (CNN), which is, otherwise, dedicated to image recognition.

REFERENCES

[1] A. Patrizio, “IDC: Expect 175 Zettabytes of Data Worldwide by 2025.” *Networkworld*. October 27, 2020. Available: https://www.networkworld.com/article/3325397/idc-expect-175-zettabytes-of-data-worldwide-by-2025.html

[2] L. Banica and M. Radulescu, “Using Big Data in the Academic Environment,” *Procedia Economics and Finance*, vol. 33, pp. 277–286, 2015.

[3] “IEEE IC Big Data Governance and Metadata Management: Standards Roadmap,” in *IEEE IC Big Data Governance and Metadata Management: Standards Roadmap*, pp. 1–62, July 3, 2020.

[4] C. K. Toh, “Security for Smart Cities,” *IET Smart Cities*, vol. 2, no. 2, pp. 95–104, July 2020.

CHAPTER 2

General Aspects Related to the Field of Big Data and Big Data in Power Engineering

2.1 BACKGROUND OF THE BIG DATA ANALYTICS

The huge volume of data mentioned in the previous chapter is basically used in three important fields: forecast, fraud detection, and business intelligence. In the past, the data warehouse was a key technology as far as business intelligence was concerned. Big Data, on the other hand, became more efficient in this domain since it can handle huge volumes of not structured or semi-structured data, whereas the data warehouse can handle only structured data (relational or nonrelational). There is also a combination of the two called Big Data Warehouse which combines the warehouse architectures with the typical Big Data technologies [1].

According to Barbara Lewis [1], the most important features of this new concept are:

the advanced analytics – besides the classical analytics methods, advanced analytical engines can be employed dedicated to predictive analytics, or even spatial and graph analytics modules developed by SAP® or Oracle® (for Hadoop).

interactive analytics – that can be performed also by human operators.

various data environments – this represents the combination between various devices for cloud storage, data storage on premises, and other hybrid environments.

data management – here the most important aspects are the removal of the outliers, data integration, format, and security as well as the data governance (for the compliance with the latest regulations).

the processing capabilities – within the Big Data Warehouse, the processing can take place at various layers. At a lower level, Hadoop could deliver aggregated information to the relational databases while at a higher one, Spark can be used directly for Big Data [1].

the repositories – destined either to Big Data or to the structured data.

Big Data appeared given the impossibility of processing volumes of petabytes, exabytes, or even terabytes of data, using a single machine. Thus, a whole series of algorithms or scalable distributed computing techniques dedicated to this have been developed. As far as these two approaches are concerned, partitioning and sampling represent two important strategies in order to speed up the data processing [2]. In this sense, the divide-and-conquer [3] plays a key role since it represents the basis for the MapReduce model which is further employed by the collection of open-source software utilities Apache Hadoop. Within this model, a Big Data file is divided into blocks, not overlapping with each other, and then partitioned among the nodes in the computer cluster using the Hadoop Distributed File System (HDFS). The distributed blocks are then processed through the two main operations: Map and Reduce.

This Apache Hadoop facilitates the finding of a solution for problems involving Big Data, using a network of many computers. Thus, it possesses the storage part, the HDFS, and a processing part represented by the MapReduce model mentioned earlier. This platform, despite the fact that it is one of the most encountered, is not very fast and, at the same time, does not perform the processing in real-time due to the high throughput latency [2]. Because of that, it is not suitable to be used in power engineering where data processing in real-time represents an important challenge. Hence, the main advantage of Apache Hadoop is that it provides a very reliable programming environment for Big Data processing [2].

Another framework dedicated to Big Data processing is Apache Spark which introduced the Resilient Distributed Dataset (RDD) [2,4] for distributed in-memory data-parallel computing. The RDDs are a fault-tolerant, unchangeable, read-only collection of objects [2] distributed among the nodes in a cluster. For these objects, two types of operations are carried out: transformations and actions. The transformations take long periods and are deterministic. Examples are mapping, filtering, and so on. The actions, instead, launch the computation on the RDD and consist of collecting, reducing, and counting [2]. Then, they store the output in a dedicated storage device.

Besides HDFS, there are also other distributed file systems that provide reliable storage solutions. Among these can be mentioned: Google File System® (GFS) or Microsoft Cosmos.

Another environment dedicated to Big Data processing is the SAP HANA®. This is in fact an in-memory relational database management system. At the same time, it is also column-oriented [5]. It serves practically as a database server and thus its most important function is data storage. The second function is to retrieve data requested by various applications. On the other hand, it can also carry out forecasts, text analytics, text search, spatial data processing, graph data processing, or streaming analytics. Hence, it can also act as an application server, and not only as a database server. That is because it possesses a more structured approach when it comes to Big Data processing. The in-memory feature permits this environment to work faster. Thus, SAP HANA can store volumes of almost 1 petabyte and respond to queries in under a second [6]. The disadvantage, in this case, is associated with the RAM costs and so the scaling-out is appropriate only for time-critical applications [6].

The library of this environment possesses various kinds of implemented algorithms that can be used in forecast, statistics, clustering, classification, and also in time series processing. The graph engine uses the Cypher Query Language that allows efficient data querying within the graph. The graph data is directly stored within the columns of the relational data.

As far as the text search is concerned, SAP HANA incorporates also a function as to how relevant the search results are as well as a threshold for the search accuracy.

As far as the deployment is concerned, SAP HANA can be installed either on the premises or in the cloud, through a number of official cloud service providers [7]. The existing hardware components can be used as well, this configuration being named Tailored Data Center Integration (TDI) [8]. In the case of on-premises installation, POWER Systems from IBM®

can be used, these being adapted to the work with AI and Machine Learning (ML) algorithms.

There is also a third possibility of efficient Big Data processing, namely SAP HANA Hadoop Integration. This combines the in-memory processing power of SAP HANA with the highly reliable Hadoop's capability to store and also to process vast volumes of data [9].

SAP HANA possesses a more structured approach when it comes to Big Data analytics while Hadoop works with large amounts of data, being dedicated more to unstructured data scenarios. Practically, this type of solution combines also the lower storage cost and flexibility of Hadoop with the in-memory and structured data conformity of SAP HANA [9].

Usually, this type of solution is adopted by those clients who already have a Hadoop cluster installed and want to use it together with SAP HANA.

SAP has developed also an environment, called SAP Vora, which runs on a Hadoop cluster [9] and Spark framework. This environment also provides in-memory processing engines, being implemented to be used in large distributed clusters and, at the same time, to scale to thousands of nodes.

Another important principle related to Big Data processing is data partitioning. Its goal is either efficient query of large databases or computation based on large volumes of data.

In 2014 and 2016, various partition methods have been proposed for Big Data stored on NoSQL data environments [10,11]. Basically, there are three types of partitioning: functional, vertical, and horizontal.

The functional partitioning is used to isolate the read-write data from the read-only data and thus improve the query efficiency within the database [2].

The vertical partitioning can be carried out through an optimal approach (taking into consideration various constraints) or heuristic approach. This consists of dividing the columns of the dataset into subsets sharing a key column. This division takes place as a function of the usage pattern [2]. At the same time, the vertical and the horizontal partitioning can be combined to divide the dataset in accordance with the workload or the target. Hence, it results in the hybrid partitioning.

The most important is horizontal partitioning because it is used by most of the Big Data applications, including Hadoop. Also, SAP uses it, despite the fact that it is column-oriented.

The horizontal partitioning uses a multitude of schemes in dividing the data. These are: composite (range-hash and range-range), Round-robin, random, hash, and range [2]. The last three are used in the field of Big Data and that is why these will be described in detail.

As far as the **random partition** is concerned, the data is organized randomly in subsets through generating a random number, based on which a certain position is designated to each record. As in the case of **Round-robin**, the generated subsets are approximately equal but unlike this method, it requires greater computational resources, given this random number generator which is in fact used for each data in the subset [2]. The main advantage, in this case, is that it provides a balance between the subsets which is compatible with the more structured approach like the one used by SAP HANA.

In **hash partitioning,** the data is partitioned in subsets by hashing the record key and mapping the key hash value to a certain partition. The most used method for the mapping is to *mod* the hash key with the number of partitions [2], in this way resulting in a partition ID. Practically, if a certain record has a key value, then the hash value must be identical. At the same time, this method does not lead to order among the partitions.

In **range partitioning**, the data is segmented according to a pre-set range over consecutive ranges [2] and guarantees the order of the partitions. The most important aspect is to define this pre-set range mentioned previously. That is because there are no dedicated methods to finding it and because these partitions are distributed among the machines in the cluster, this range has to be very precisely calculated. This means further supplementary costs and thus a compromise solution between costs and accuracy has to be determined [2].

If one has to draw a conclusion about these three subtypes of partitioning, one can say that the random partitioning carries out a sequential scan of the entire database and assures a good data partition balance. At the same, time it requires additional computing power in order to generate the random values. The data is organized also randomly and there is no order of partitions. On the other hand, the hash partitioning carries out also a sequential scan of the database and is very efficient when conducting point queries [2] based on partitioning attributes. This means that, in this case, only one partitioned has to be searched. The hash partitioning is not appropriate neither for range queries nor for point queries on non-partitioning attributes [2].

Finally, the range partitioning, like the previous two methods, carries out a sequential scan of the entire database. It generates a good balance of the data partition and is suitable for range and point queries. It is also similar to hash partitioning in that it requires the searching of only one or few partitions. Unfortunately, execution skew can arise given the data processing in one or more partitions. For instance, this is not the case for hash partitioning where only one partition is searched.

The basic idea of data partitioning is to obtain optimum scalability among the computers in a cluster when it comes to processing vast volumes of data. Hence, this principle is of paramount importance for the field of Big Data.

As already stated, in the case of Hadoop clusters, the partitioning is done through HDFS. When a file is imported into HDFS, this is segmented into blocks of a fixed size [2]. In Apache Spark, for an improved querying efficiency, the first stage is to import these HDFS blocks into in-memory RDD. This can be further partitioned using any of the three partitioning subtypes already mentioned. Not having a structured approach, Hadoop does not take into account the statistics of the data in partitioning it. Another major disadvantage is represented by the data skew. This is the uneven data distribution that leads implicitly to execution skew (different execution times for the tasks).

It is important to know that within the computer clusters, performance is strongly influenced by data partitioning and thus by the evenness degree of data distribution among the cluster components. This counts not only for the mapping operation but also for the reduced operation. The data evenness corresponding to the mapping operation is a function of the unbalance in the input data. That is why the time series serving as input data for the load forecast have to have the same dimension. On the other hand, the data evenness corresponding to the Reduce operation is a function of the unbalance in the intermediate data [12,13].

This data unbalance influences as well the efficiency of the ML algorithms. As a result, the random partition has to be employed as it assures the data evenness across the nodes in the cluster [2]. A promising application in this direction, dedicated to Hadoop, is the Cloud OnLine Aggregation (COLA).

When it comes to Big Data, in order to decrease the computational burden, sampling plays a key role. On the other hand, sampling helps also to better consider the statistical aspects of the data that, usually, are not considered by partitioning (especially on Hadoop clusters). This leads automatically to increased inefficiency of Big Data processing.

Data sampling is practically always used in combination with partitioning. When it comes to sampling, the data quality rather than quantity is of great importance. Based on this procedure, a better assessment of the partitioning can be obtained and, implicitly, an improvement in data processing times can be attained.

There are many sampling methods but random sampling is of great importance for Big Data. That is also because together with various sampling methods a certain bias in the data can result. That is why this random sampling is, probably, better suited from this point of view. But if one decides that better samples should be used, other techniques (different from random) must be employed. So, here, again a compromise solution has to be found between obtaining a smaller bias and using better samples. Given the fact that is easy to select nonrepresentative data, when working with Big Data, one approach is to use as much data as possible and to run the experiment at scale [2].

Following, the most important sampling techniques are described. All of them can be used together with Hadoop. As far as SAP HANA is concerned, this can be used with all, except the Reservoir sampling.

Reservoir sampling is carried out without replacing data from a vector. The length of this vector is either unbounded or indeterminate [14]. The data is received from this array (vector) and a sample is being maintained in a buffer called reservoir.

Reservoir sampling is recommended when the input data possess an unknown number of records or when there are too many records to be stored. In both cases, random sampling does not work. The Reservoir sampling delivers good results for large datasets.

Random sampling, mentioned earlier, is the most flexible, most general, and most encountered method. Within this procedure, each item in the dataset has virtually the same probability to appear in the sample. Depending on the use of the replacement, an item can appear in the sample either once or more than once. As the name of the method tells, one cannot be assured of the fact that the items that appear in the sample are also the most representative. That is why **Stratified sampling** must be used in this situation.

Within the Stratified sampling, the dataset is divided into a disjoint set of data groups called strata [2]. From each stratum, a random sample is extracted and each of these extracted samples is then combined in order to generate the sample. It is considered that the records in each stratum is representative and that no stratum is neglected. The main advantage of this sampling method is that it requires a smaller sample size than random sampling in order to obtain an identical accuracy with Random sampling [2].

Within **Bernoulli sampling**, on the other hand, each record in the dataset possesses the same probability to appear in the sample [15]. There is no replacement and each record is selected independently for the sample [2], meaning that the sample size is random.

In the case of **Bootstrap**, the variability of the sample is assessed. It demands high computation power because the size of the extracted sample is in the same order as the initial Big Data set and the entire processing is carried out on this sample.

As already mentioned, any of these methods can be employed together with Hadoop. The configuration of a Hadoop cluster is based on blocks. Thus, the sampling can take place either at the block or at record level.

In the first situation, the dataset is stored as disjoint blocks and each block contains a small subset of items. When the sampling procedure is carried out, an entire block is selected to become part of the sample. This kind of approach is very efficient since not only the storage but also the processing itself are organized on blocks. At the same time, another advantage is that in this case, one needs lesser block accesses [2] than for the record-level sampling. If the blocks are correlated, the results can be biased. The same happens also with the RDDs mentioned previously. This problem is solved through the Random Sample Partition (RSP) proposed in Ref. [16]. That is because this method transforms the HDFS blocks into random sample data blocks that can be used without the need for further processing. Possessing highly reliable statistical estimators, the RSPs can be employed directly in forecasting or in any other statistical determination.

The record-level sampling is highly dependent on the chosen individual records from the input data. This type of sampling passes through all data items sequentially and that is why the processing time is very long. The main conclusion arising from this situation is that if the HDFS is too large, this sampling type passes through each record and thus is not efficient. If Bootstrap is used in this context, the whole procedure becomes even more time-consuming (because, as already mentioned , the samples, in this case, have a comparable size to the initial Big Data set). A possible approach to reducing the size of the samples could be either the Bag of Little Bootstraps (BLB) or the Divide and Conquer algorithm [2].

What is more important is that sampling can also be an efficient approach in solving the data skew problem [2].

Another promising solution dedicated to the reduction of computer resources is the so-called approximate computing. This allows also a reduction in the consumed electrical energy [2]. This type of environment is used on a small synopsis of data instead of the entire data set. In most situations, when working with Big Data is highly probable to obtain approximate results and not exact values. That is because the Big Data set can contain outliers or values generated as a result of communication faults. Given also the fact that approximate computing is much more efficient in terms of processing time, one can say that it represents a compromise solution between speed and accuracy.

The approximate computing can be combined with incremental computing [2]. In incremental computing, when a data item changes, only those outputs are recomputed which depend on the changed item. Thus, the processing time is shortened. Within the combination between these two procedures, the results obtained from the Big Data processing are updated in accordance with the incremental computing techniques.

Despite the fact that there many approaches dedicated to Big Data storage and processing, like histograms, sketching, aggregation and so on, sampling is the most efficient and most encountered.

Approximate computing is used, practically, only when approximate results are enough. There are instruments for approximate computing, dedicated primarily to work with Hadoop, like: ApproxHadoop, ApproxSpark, IncApprox, BlinkDB, BlinkML,

EARL, ApproxIoT, Sapprox, or the newest method proposed in [16], the RSP.

ApproxHadoop uses multistage sampling and hence the MapReduce procedure becomes more efficient in terms of computing time.

ApproxSpark, like the ApproxHadoop, can use multistage sampling or even Reservoir sampling in order to carry out the approximation in Apache Spark. Although it can work either with block-level sampling or partition-level sampling, block-level sampling causes greater errors due to the fact that the data in the RDD is not randomized [2].

IncApprox is stream-data analytics and is based on incremental computing for updating the approximate results.

BlinkDB is a distributed query engine based on sampling.

BlinkML is an approximate machine learning technique, working on a small sample instead of the entire Big Data set. Its main advantage is that it provides error limits for the accuracy of the approximate model. The error limits are estimated, based on linear regression, max entropy classifier, Probabilistic Principal Component Analysis (PPCA), logistic regression, or maximum likelihood estimation [2]. The sampling used by BlinkML is uniform or it can use also Bernoulli.

Early Accurate Result Library (EARL) represents an extension of Hadoop and provides early estimation results [2]. Like BlinkML, it uses also a uniform sampling procedure from the HDFS as well as Bootstrap to evaluate the accuracy [2].

The ApproxIoT depends on the edge computing capabilities to carry out the approximation based on a sampling of the data resulting from the IoT sensors. The sampling procedure is the reservoir.

The Sapprox collects the data subsets offline and uses these to facilitate the online sampling. It further uses cluster sampling with unequal probability to solve the data skew problem [2].

The RSP approach is described in Ref. [16].

Mention should be made of the fact that the MapReduce procedure is efficient only when the entire Big Data set is scanned only once. For multiple iterations of this kind, it is completely ineffective. That is because the access, in his case, is made through a disk. On the other hand, the Apache Spark possessing an in-memory computing model is much faster and can cope also with the problem of iteratively scanning the entire dataset provided that the memory is large enough to store all the data blocks.

Regarding the partitioning, a huge improvement would be the development of methods that are able to take into account also the statistics related to the datasets. Another issue is that in the present times, many applications work with distributed and unbounded data. The extraction of a random sample from these data is always very difficult due to the fact that when the data is unbounded one cannot determine the sampling probability. The data distribution among the computer in the cluster makes unfeasible the collection of the data on a single machine, for sampling. At the same time, also online sampling proves very difficult. That is why the partitioning of the Big Data set into smaller subsets plays a key role when it comes to Big Data processing.

From the current trend, one can see very easily that most of the applications use the so-called batch (data in rest) processing. That is data that does not move from device to device or network to network, like, for instance, data stored on the disk drive, flash drive, or data that is archived or already stored in some other way.

On the other hand, most of the time, the input data for an application cannot be stored, given its volume. Thus, the question arises whether the entire data needs really be stored or is it enough to use only a subset? The obvious solution could be incremental approximate computing [2].

Another approach could be a combination between approximate and cluster computing, in which accuracy, aggregation, sampling, and selection are of great importance.

2.2 THE BIG DATA AND THE POWER SYSTEM: SOME GENERAL ASPECTS

The attributes of Big Data in general, but also in the power system is characterized by the five Vs:

Volume: Is characterized by values of at least TB. For the power system, it can be even less, with an order size of GB.

Velocity: In order to take the appropriate business decision, the data have to be available at the right time. This is valid also for the power system in which the data streaming takes place in real-time. This is one of the most important challenges when handling Big Data in power systems because, when done properly, it can assure efficient dynamic pricing of the energy.

Variety: Data comes from all kinds of formats, more or less structured.

Veracity: This concept is related to data quality. On one hand, the data can contain outliers or values generated by communication errors. On the other hand, they could be or not be relevant for the problem in discussion.

Value: If the data is relevant, these can be transformed into business and implicitly in income.

Usually, Big Data is used either in forecasts, for user behavior analytics, fraud detection, and so on. The most important challenges in the field of Big Data are collection, storage, analysis, processing, visualization, or even information privacy [17].

When it comes to power systems, the concept of Smart Grid (SG) is of great importance. In the end, it assures efficient energy consumption based on bidirectional communication infrastructures like advanced metering infrastructure (AMI) and wide-area monitoring systems (WAMS). Both of them generate enormous volumes of data. For example, in the USA, the smart meters generate approximately 1000 petabytes per year. The size order of the data volume is represented by terabytes. When considering the transmission system and the distribution system as two separate entities, in principle the transmission system generates less data than the distribution system given the presence of smart meters within the last one. Even a household can generate data volumes in the range of GB per year, given the presence of the same meters. This order size of GB qualifies as Big Data only for the power system paradigm.

Coming back to the SG, this possesses various data sources like field measurements obtained by various substation/feeder smart devices, specialized commercial, state (or government) databases related to weather, seismic data, lightning, electricity market, vegetation, soil data, and so on.

Thus, energy consumption can be addressed with specific Big Data techniques given also a large number of communication devices, within the communication infrastructure of the SG, that generate sometimes partially structured data or even unstructured data. As it was mentioned in the previous section, this is another characteristic of Big Data. On the other hand, also the presence of the five V's that characterize the Big Data, can be observed also in the data generated by the power system.

When it comes to SG, a very important parameter is the demand response (DR). This requires an accurate load forecast and especially peak prediction [17].

On the other hand, the transmission and distribution operators within an SG rely on data analytics in order to identify the potential anomalies in the power delivery, outages, and implicitly to restore the service faster.

Another key aspect is that even if the data volume is not always big, the real-time processing requirements pose identical challenges to those posed by Big Data [17].

In the end, the supreme goal is to have online demand response algorithms which manage this parameter in real-time. Of course, this is very hard to achieve, especially on a large scale.

Thus, one is forced to rely on either price or load forecast that becomes highly important in this situation.

The load forecast is based, among other elements, also on exogenous variables like weather or even social networks, electricity markets, etc. [17].

Another important aspect to be taken into consideration is that even if the utility company has already measured the delivered power on the generator's side, the measurements on the customer's side are more relevant and precise given the fact that transmission and distribution losses are also included [17].

In order to carry out such a load forecast, one has many methods at the disposal. The simplest one and the most encountered is the Linear Regression. But, as one shall see in Chapter 3, the Convolutional Neural Networks (CNNs) are very accurate and show very promising results. At the same time, this method is also very compatible with the concept of Big Data.

Within this chapter, in the sequel, one shall present a simple method of implementation for the Linear Regression in order to have also a comparison with the CNNs and also because within the research and industrial sectors, Linear regression is the most utilized.

As far as the regression analysis is concerned, the most straightforward method is to consider the energy consumption y and the temperature x. Due to the fact that energy consumption is always influenced by temperature, within this analysis also conditional expectations have to be considered.

Basically, if we want to predict the value of y, considering that it is not influenced by any other variable, its expected value is described in equation (2.1):

$$E(y) = \int yf(y)dy \tag{2.1}$$

This expected value represents a minimum mean square error predictor. If p is the predicted value, then the mean square error is given in (2.2):

$$M = \int (y - p)^2 f(y)\,dy = E\{(y - p)^2\} = E(y^2) - 2pE(y) + p^2 \tag{2.2}$$

It is obvious that if $p = E(y)$, the value given in (2.2) is minimal.

When y is related to x and x is already known, then the joint distribution of these two variables can be denoted as $f(x, y)$. Thus, the new minimum mean square error is described in (2.3):

$$E(y|x) = \int y \frac{f(x, y)}{f(x)} dy \tag{2.3}$$

where:

$$f(x) = \int f(x, y)\,dy \tag{2.4}$$

In (2.4), the marginal distribution of x is described.

The conditional expectation can be described also based on equations (2.5) and (2.6):

$$\hat{y} = \hat{y}(x) = E(y|x) \tag{2.5}$$

$$E\{(y - \hat{y})^2\} \leq E\{(y - p(x))^2\} \tag{2.6}$$

At the same time, the expectation value $E(xy)$ is described in (2.7).

$$E(xy) = \iint xyf(x, y)\partial y \partial x = \int x\{\int yf(y|x)\partial y\} f(x)\partial x = E(x\hat{y}) \tag{2.7}$$

In our case, x and y are normally distributed. This means that $E(y|x)$ is described as in (2.8):

$$E(y|x) = ax + b \tag{2.8}$$

In fact, (2.8) represents the linear regression equation. The idea is to find the values for a and b as a function of E(x), E(y), the variances V(x), and V(y) as well as the covariance C(x, y).

The equation (2.8) can be also written as in (2.9):

$$E(y) = aE(x) + b \tag{2.9}$$

Thus, the value for b can be obtained, based on (2.10).

$$b = E(y) - aE(x) \tag{2.10}$$

Hence, the expected value $E(y|x)$ can be described also as in (2.11).

$$E(y|x) = E(y) + a\{x - E(x)\} \tag{2.11}$$

In other words, the difference between the conditional expected value of y and its unconditional expected value is directly proportional to the difference between the error of predicting x and its expected value.

The next step is to multiply (2.8) [18] by x and $f(x)$, integrate it with respect to x and obtain $E(xy)$ as described in (2.12).

$$E(xy) = bE(x) + aE(x^2) \tag{2.12}$$

At the same time, the product of the two expected values of x and y is computed in (2.13).

$$E(x)E(y) = bE(x) + a\{E(x)\}^2 \tag{2.13}$$

Subtracting (2.13) from (2.12), a is obtained, according to (2.14).

$$a = \frac{E(xy) - E(x)E(y)}{E(x^2) - \{E(x)\}^2} = \frac{C(x, y)}{V(x)} \tag{2.14}$$

At this moment, a and b have been determined and thus the load forecast through linear regression is carried out. The problem is that this kind of approach does not take into consideration multiple conditions. The load forecast can be influenced by other factors related to weather, not only the temperature, as well as by market factors like, for instance, the energy price. That is why, more realistic mathematical models are needed when determining the load forecast, especially for longer periods of time.

Besides load forecast, Big Data processing techniques can be used in fraud detection as well. This can be, basically, of three types: an attack on the data of the customers, reporting less energy consumed than actually used, and more energy consumption through rogue connections than actually declared [17]. Each of these types can cause perturbances as far as the grid control is concerned.

In the first instance, the client privacy can be compromised; in the second, the demand response will send a fake energy requirement to the generators; while in the third, if the SG is connected to a power plant, this can begin to generate more electricity than is actually required and the energy price will increase [17].

The SG possesses various types of sensors, especially along the power lines. The attack on the data resulting from these sensors is the most dangerous because it can lead to potential blackouts. Another source of data, besides the sensors mentioned earlier, can be also the Phasor Measurement Units (PMUs). These generate high volumes of data at high frequency, meaning 60–120 frames per second for 60 Hz systems. Another piece of equipment that still is of paramount importance when it comes to grid control is Supervisory Control And Data Acquisition

(SCADA). A SCADA system together with unsynchronized measurements resulting from Remote Terminal Units (RTUs), the various line sensors, as well as PMUs are very useful to state estimation (SE). Through SE, problems of grid operation in real-time can be determined. That is why the potential of Big Data also in this direction is enormous.

The SE can be conducted either iteratively or non-iteratively. The first type is based on power flow equations. In this case, one method to speed up the computation is the DC power flow model coupled with linear regression. The second type [19] can be based only on PMUs, or SCADA, or a combination between the two. But for this, the power system must be observed with sufficient redundancy by the PMUs. Other non-iteratively methods have been proposed in [20,21].

Practically, Big Data in power systems aims at improving the grid analysis for efficient consumption and, at the same time, electricity generation. It also facilitates load management, correlating it better with the actual demand. From this, a whole series of economic benefits result not only for the customer but also for the electricity producer.

So, the first conclusion that can be drawn is that Big Data is very useful to market and of course pricing when it comes to power systems. To this goal, in theory, both transmission and distribution data have to be analyzed. In practice, that would be easier only where both the transmission and distribution are managed by the same company. But this is a rare case, especially in Europe. This is the first challenge. There are also others, as they will be described in Chapter 4, dedicated to conclusions and future research directions.

2.3 THE BIG DATA IN THE POWER SYSTEM: SOME CHARACTERISTIC FEATURES ABOUT STORAGE AND ANALYTICS

The Big Data analytics applied in power systems is not a homogeneous process. It supposes the use of multiple concepts from various disciplines that are not limited to the analytics itself but are related as well to the relevance of these data from the point of view of the customer and/or the producer.

Another point to tackle is that in power engineering, the problems are often solved based on models. In the IT sector, the solution has more of an exploratory character and is based purely on data. That is why a possible approach in solving the problems in power engineering through Big Data is to use a combination of the two paradigms [22].

The general processes in which Big Data is used are based on diagnostic, prescriptive, corrective, predictive, or even descriptive measures.

The diagnostic procedures can be easily correlated with the data mining. Thus, the classification or clusterization is dedicated to attributing a certain feature to the classes or clusters that resulted from these operations. Here, MapReduce or SQL represents valuable algorithms. Other tools that play a key role within the distribution grid are DISTIL or BTrDB because allow the fast processing of vast volumes of data resulting from the measuring equipment like the microPMU. Both these tools are devoted to rapid statistical queries and fast difference computation [22]. This permits further detecting the transient within the network. Current studies in this direction are aimed at facilitating online event detection and localization as well as determining the phase imbalance within a distribution network [23].

The prescriptive and corrective measures can be correlated with the statistical algorithms applied to databases. Unfortunately, this means selecting only a portion of data that is to be treated with R, MATLAB®, SAS®, and so on. And this means further sampling. In this sense, one can use any of the methods described in section 2.1. Another approach is to combine the statistical processing with the database. A good choice can be ScaLAPACK [22].

The predictive measures are intrinsically related to massive parallelism. The idea of massive parallelism comes from the fact that within the power grid there is a plurality of data sources like AMI, PMUs, or microPMUs. Data patterns for decision making and also for the prediction can be obtained from the aggregation or disaggregation of the flows resulting from these data sources. This is extremely useful above all in the case of SG possessing distributed energy sources like solar, wind, and so on. In this way, a disaggregated prediction for the load forecast can be carried out.

Such an application dedicated to disaggregating data flows is based on Non-Intrusive Load Monitoring (NILM). This method refers to the analysis of currents and voltages in a house and deducing not only what appliances are used but also what is their energy consumption [24]. Based on this principle, at the grid level, using a clusterization algorithm, the different types of appliances can be classified into various categories, and in this way, a very precise behavior of the client can be predicted.

The descriptive methods, on the other hand, are related to complex event processing [22]. This type of processing is strongly compatible with Big Data as it takes place in real-time. It often refers to stability or power quality.

Another important aspect that has to be taken into consideration is the so-called models that scale up. These were initially destined to work with problems whose solution did not rely on Big Data. This means that now these methods have to be integrated together with algorithms that are capable of processing vast volumes of data [22]. This was especially the case of the nonlinear AC optimal power flow.

Despite the fact that Big Data in power systems do not possess the same characteristics as Big Data in the IT sector, especially in terms of volume, many energy producers and regulatory organisms identify this novel concept as essential in modernizing the power grid at every level. Among these organizations, one can find the US Department of Energy that states that by 2030 an increase from the current 1% to approximately 14%, at the national level, in solar energy production has to take place [22].

In spite of the fact that the basics of the infrastructure exist, as far as the SG is concerned, the data collection, storage, processing, and communication within the grid must suffer a revolution in order to employ, at a large scale, this concept. Thus, for this revolution to take place, one has to cope with multiple problems and challenges.

The first one and most important is real-time processing. Depending also on the consumer type, many applications in the power grid require real-time collection and processing. These take place slightly differently as in the IT sector since the power grid represents a critical infrastructure. Above all, these two operations (collection and processing) influence decisively the transmission and distribution of electrical energy [22]. This is further related to the fact that real-time processing can improve the operation of many of the distributed energy sources that have a volatile character as far as the generation is concerned. A common issue here is when, for example, the solar irradiance is high but there is no consumption. In this case, the data processing can trigger battery storage at the right time.

On the other hand, Big Data analytics improves the circulation of voltages and currents resulting from renewables, a common problem being the back injection of the power flow. But, again, the communication equipment is of great importance. In the majority of cases, this communication is not reliable and the beneficiary, being not aware of its importance, is not willing to invest too much. In this situation, the problem is not only the speed but also the bandwidth. A possible solution is the adoption of the Long-Term Evolution (LTE) standard which is used in the mobile telephony sector as well. This standard is dedicated to broadband and rapid communication even between data terminals. The main characteristics of this standard [25] are improved spectral efficiency, scalable bandwidth, it can support many types of users, and increased downlink and up-link peak data rates. Other advantages are improved system capacity and coverage, high data rate with reduced latency, and reduced cost for the operator.

From the principle point of view, the data management should take into consideration that data destined to be exchanged in real-time with the distributed energy resources within the SG must not be characterized by large volumes. Here, the predictive SE plays a vital role. Another aspect one has to be aware of, is that the cybersecurity applications that protect that data communication will lead to increased latency.

The next challenge is represented by the fact that within the power system, the centralized and distributed data management must coexist [22]. That is because the old central data management systems will not disappear and cannot be completely transformed. On the other hand, the customers can be distributed over a wide geographical area. Probably, in the more distant future, utility companies should contribute to the development of Big Data management systems dedicated to power engineering.

In some cases, given the fact that Big Data in power engineering is directly correlated with consumer taxation, the real-time requirements for data processing can overcome the same requirements in the IT sector. In this situation, the classical approaches based on Hadoop, SQL, or NoSQL are not sufficient and the various equipment manufacturers should contribute also to the development of dedicated Big Data real-time processing platforms like, for instance, the BTrDB, mentioned previously. This environment is able to sustain queries and analysis for an expected 44 quadrillion data points per year and per server [22]. The technology used by BTrDB permits scalable analysis in spite of the asynchronous changes in data which are very often encountered in the field of power engineering.

Another challenge is the cyber-security applications which are very important not only for communication but also for the stored data. Must these applications be managed in a centralized manner or dispersed? Probably a combination between the two is most appropriate.

The next important aspect is represented by the siloed data [22]. The situation is, obviously, very different from country to country due to the various regulations. In some European states, indicators pertaining to blackouts or to what caused them, are not public. At the same time, as already mentioned, if the transmission and distribution are not managed by the same company, the two entities will not have access to data from the other party. Thus, this concept is not easy to tackle. Some of the states argue that after the BlackEnergy malware interrupted various consumers in Ukraine in 2015, that siloed data is necessary. This is the first cyberattack on a power grid that has been acknowledged as successful. The malware was introduced by phishing emails. In a first instance, the SCADA control was seized, interrupting remotely various substations. In the end, also a denial-of-service attack on the call center has been carried out, preventing the customers to be updated on the status of the blackout. This malware can be removed by a special dedicated tool named YARA. Until these issues are not fully and reliably addressed, the data siloes will remain in place and this could prove an obstacle in adopting Big Data analytics for customers residing over a large geographical area.

The next challenge is to address discarded data [22]. In the past, only the data with an immediate value have been employed for monitoring and overseeing purposes. Big Data means also the use of unstructured data that in a first instance does not possess a clear purpose. The main idea here to develop various applications which can prove in time very useful for the power grid. This means, in other words, to develop applications that use the data discarded in past to extract different patterns in order to identify outages within the network, models to use various renewable sources, or even to try to extract information regarding the fatigue degree and wear of various equipment in the grid [22].

It is useful to have also a better idea about the types of processing algorithms that can be used for Big Data in the power grid. These are ML algorithms and can be classified, as usual, in algorithms for Unsupervised Learning, Supervised Learning, and Reinforcement Learning. The first category can be associated with clustering, and dimensional reduction; the second to either classification or regression; the third to those algorithms that react to the environment [26].

Coming back to the first category, mention should be made of the fact that this uses data that is not classified in any way. The goal is to search for previous undetected patterns with minimum human supervision One of the most important applications of unsupervised learning is density estimation in the field of statistics.

Supervised learning, on the other hand, is carried out based on historical data. That is why this type of ML algorithm is best suited when it comes to load forecast. Another reason in the same direction is that being directly influenced by a human operator, the training set, as well as the learning function or algorithm, are chosen specifically for the problem. In the end, the same human operator, after training, must evaluate the accuracy of the results against a test set, completely different from the training data set.

Reinforcement learning consists of various methods in which a software agent learns a certain strategy in order to maximize his reward. The basic idea, in this case, is to find the right balance between the exploration of an uncharted area in the data set and exploitation of the current knowledge in the data set [27].

Another important aspect is to put in relation the 5 Vs mentioned above with the specific corresponding dimension in the power systems. The most important difference concerns the volume. That is because, in power engineering, Big Data does not mean necessarily the size order of petabytes and exabytes. For example, a PMU generates on average 30 GB of data a day. A smart meter 120 GB [26]. If one wants to combine these data with weather data, the situation does not change radically. Thus if a radar is being used, this will produce approximately 612 MB per radar scan. Of course, it depends on how many radar scans have to be used. From a weather satellite, at least 10 GB of data a day result [26]. As a rule, if a weather forecast model is employed, this will generate between 5–10 GB a day of data.

If aspects related to topography and vegetation have to be taken into consideration, this will add another 2.7 GB of data per day. Of course, all these values are approximate but, in spite of this, they do provide an idea about the data volume in a power grid, based on the complexity of the model one wants to adopt. So a very important conclusion arises, this being the fact that in a distribution system the data quantity is larger than in the transmission system.

Now, what about the velocity? For the sake of simplicity, one will consider the same data categories. Thus, the data coming from the PMU are characterized by 240 samples/sec while in the case of the smart meter, a measured value is transmitted every 5–15 minutes [26]. These are the most encountered parameters. In special applications, one can have also 1 minute or even less. But, usually, if one refers to the transmission system, the key parameter is one value every 15 minutes (as generated by the SCADA system).

As far as the weather data is concerned, the radar produces a sample every 4–10 minutes, the satellite every 1–15 minutes, the Automated Surface Observing System (ASOS) every 1 minute while the data resulting vegetation and topography is static, from this point of view.

Regarding the variety, one has three to four sources for each data category. Hence, the power data comes mainly from smart meters, IoT devices and sensors, or digital fault recorders (DFR). On the other hand, the weather data are generated by ASOS, satellite, or radar. The vegetation and topography data result in general from a Light Detection and Ranging (LIDAR), or more specifically, from the Ecological Mapping System of Texas (only for this area) [26].

Regarding the data veracity, mention should be made of the fact that PMU data error is less than 1%. In the case of smart meter, this parameter is less than 2.5 % and in the case of the DFR is even lower than 0.2% [26]. For the data coming from the radar, the noise is less than 1.2 dB/ms.

When storing and processing Big Data within the power grid, cloud computing turns out to be an interesting approach. It enables on-demand access to a multitude of computer resources. This approach offers various service models like Platform as a Service (PaaS), Infrastructure as a Service (IaaS), Software as a Service (SaaS), or Network as a Service (NaaS) [17].

In PaaS, the service provider controls the cloud infrastructure. The beneficiary has at his disposal the capability [17] of developing software applications using tools for which the access is granted by the provider.

In the case of the IaaS, the service provider manages the physical infrastructure of the cloud. The client has the possibility of controlling the storage as well as some of the processing resources. At the same time, the customer is able to run arbitrary software applications [17].

For SaaS, the service provider manages the servers, operating systems, and other applications pertaining to the cloud. It is the easiest and most comfortable type of service since the client access to the cloud takes place through an online dedicated interface.

In NaaS, the service provider puts at the client's disposal virtual network services through the Internet and the payment is made on the basic pay-per-use monthly subscription.

The prices vary between 0.61 $/hour for a 2 Cores processor, 28 GB RAM, 384 GB storage infrastructure, and 8.78 $/hour for an infrastructure of a 32 Cores processor, 448 GB of RAM, and 6144 GB of storage [28]. This configuration works very well together with SAP HANA, DataZen, SQL Server, Hadoop, and Hortonworks [28].

The main conclusion that arises from this chapter is that the tools specific for Big Data storage and processing, like the one mentioned in the previous paragraph, are not always needed when it comes to power engineering. They are efficient only where the Big Data volume is comparable to the IT sector. That happens above all in the distribution or even in the case of a household. But for power transmission systems, these are not the best suited since the data volume is much lower and such infrastructures incur high costs, especially when taking into consideration a longer period of time.

REFERENCES

[1] Barbara Lewis, "Building the Big Data Warehouse: Part 2." *Digitalistmag*. March 13, 2018. [Online]. Available: https://www.digitalistmag.com/cio-knowledge/2018/03/13/building-big-data-warehouse-part-2-05964490/

[2] M. S. Mahmud, J. Z. Huang, S. Salloum, T. Z. Emara and K. Sadatdiynov, "A Survey of Data Partitioning and Sampling Methods to Support Big Data Analysis," *Big Data Mining and Analytics*, vol. 3, no. 2, pp. 85–101, June 2020.

[3] N. Lazar, "The Big Picture: Divide and Combine to Conquer Big Data," *Chance*, vol. 31, no.1, pp. 57–59, 2018.

[4] M. Zaharia et al., "Resilient dIstributed Datasets: A Fault-Tolerant Abstraction for In-Memory Cluster Computing," in *Proceedings of the 9th USENIX Conference on Networked Systems Design and Implementation* (NSDI'12), San Jose, CA, USA, 2012, p. 2.

[5] Jeff Kelly, "Primer on SAP HANA." *Wikibon*. July 12, 2013. [Online]. Available: http://wikibon.org/wiki/v/Primer_on_SAP_HANA

[6] Felix Weber, "SAP HANA and Big Data – Scale-out Options." Felixweber. April 17, 2017. [Online]. Available: https://felixweber.me/2017/04/sap-hana-and-big-data-scale-out-options/

[7] SAP SE, "Planning Your Cloud ERP Deployment." SAP SE.

[Online]. Available: https://insights.sap.com/cloud-erp-deployment-options/

[8] SAP SE, "SAP HANA Tailored Data Center Integration-Overview." SAP SE. Available: https://www.sap.com/documents/2017/09/e6519450-d47c-0010-82c7-eda71af511fa.html

[9] SAP SE, "SAP HANA Hadoop Integration." SAP SE. [Online]. Available: https://help.sap.com/viewer/c618c36d79e34077a680bac61affc8b7/2.0.3.5/en-US

[10] S. P. Phansalkar and A. R. Dani, "Transaction Aware Vertical Partitioning of Database (TAVDP) for Responsive OLTP Applications in Cloud Data Stores," *Journal of Theoretical and Applied Information Technology*, vol. 59, no. 1, pp. 73–81, 2014.

[11] J. Kamal, M. Murshed and R. Buyya, "Workload-Aware Incremental Repartitioning of Shared-Nothing Distributed Databases for Scalable OLTP Applications," *Future Generation Computer Systems*, vol. 56, pp. 421–435, 2016.

[12] Z. Tang, W. Lv, K. Li and K. Li, "An Intermediate Data Partition Algorithm for Skew Mitigation in Spark Computing Environment," in *IEEE Transactions on Cloud Computing*, doi: 10.1109/TCC.2018.2878838.

[13] Y. Kwon et al., "A Study of Skew in MapReduce Applications," in *The 5th Open Cirrus Summit*, Moskow, Russia, 2011, pp. 1–5.

[14] J. S. Vitter, "Random Sampling with a Reservoir," *ACM Transactions on Mathematical Software*, vol. 11, no. 1, pp. 37–57, 1985.

[15] C. T. Fan, "Development of Sampling Plans by Using Sequential (Item by Item) Selection Techniques and Digital Computers," *Publications of the American Statistical Association*, vol. 57, no. 298, pp. 387–402, 1962.

[16] S. Salloum, J. Z. Huang and Y. He, "Random Sample Partition: A Distributed Data Model for Big Data Analysis," *IEEE Transactions on Industrial Informatics*, vol. 15, no. 11, pp. 5846–5854, Nov. 2019.

[17] Feng Ye, Yi Qian and Rose Qingyang Hu, "Big Data Analytics and Cloud Computing in the Smart Grid," in *Smart Grid Communication Infrastructures: Big Data, Cloud Computing, and Security*, IEEE, 2017, pp. 171–185.

[18] University of Leicester. Conditional Expectations and Regression Analysis. [Online]. Available: https://www.le.ac.uk/users/dsgp1/COURSES/THIRDMET/MYLECTURES/1REGRESS.pdf. 2007.

[19] A. S. Dobakhshari, S. Azizi, M. Paolone and V. Terzija, "Ultra Fast Linear State Estimation Utilizing SCADA Measurements," *IEEE Transactions on Power Systems*, vol. 34, no. 4, pp. 2622–2631, July 2019.

[20] B. Fardanesh, "Direct Non-Iterative Power System State Solution and Estimation," in Power and Energy Society General Meeting, 2012 IEEE. IEEE, 2012, pp. 1–6.

[21] X. T. Jiang, B. Fardanesh, D. Maragal, G. Stefopoulos, J. H. Chow and M. Razanousky, "Improving Performance of the Non-Iterative Direct State Estimation Method," in Power and Energy Conference at Illinois (PECI), 2014. IEEE, 2014, pp. 1–6.

[22] Hossein Akhavan-Hejazi and Hamed Mohsenian-Rad, "Power Systems Big Data Analytics: An Assessment of Paradigm Shift Barriers and Prospects," *Energy Reports*, vol. 4, pp. 91–100, February 2018.

[23] O. Ardakanian et al., "Event Detection and Localization in Distribution Grids with Phasor Measurement Units". arXiv, Nov. 2016. [Online]. Available: https://arxiv.org/abs/1611.04653

[24] G. W. Hart, "Nonintrusive Appliance Load Monitoring," *Proceedings of the IEEE*, vol. 80, no. 12, pp. 1870–1891, December 1992.

[25] LTE Encyclopedia. [Online]. Available: https://sites.google.com/site/lteencyclopedia/home

[26] R. V. Angadi, P. S. Venkataramu and S. B. Daram, "Role of Big Data Analytics in Power System Application," presented at 2nd International Conference on Design and Manufacturing Aspects for Sustainable Energy (ICMED 2020), Hyderabad, India, July 10-12, 2020.

[27] L. P. Kaelbling, M. L. Littman and A. W. Moore, "Reinforcement Learning: A Survey," *Journal of Artificial Intelligence Research*, vol. 4, pp. 237–285, 1996.

[28] Microsoft Corporation, "Cloud Services Pricing." Microsoft Corporation. [Online]. Available: https://azure.microsoft.com/en-us/pricing/details/cloud-services/?&ef_id=Cj0KCQiArvX_BRCyARIsAKsnTxOfXK4zVIHfTZMBHDV-XbDgEwJV5ttGXJ16qogOnR2UVwMoz1s9Lo8aArnsEALw_wcB:G:s&OCID=AID2100644_SEM_Cj0KCQiArvX_BRCyARIsAKsnTxOfXK4zVIHfTZMBHDV-XbDgEwJV5ttGXJ16qogOnR2UVwMoz1s9Lo8aArnsEALw_wcB:G:s

CHAPTER 3

Big Data in Power Load Forecast

3.1 BIG DATA IN POWER LOAD FORECAST AT DISTRIBUTION LEVEL

The potential of Big Data analytics is huge and the previsioned market evolution for this, till 2025, can be seen in Figure 3.1.

As can be observed in this figure, in 2011 the market volume was only 7.2 billion USD, at the global level. The growing trend was kept till 2013. Then in 2014, a first decrease appears from 19.6, in the previous year, to 18.3 billion USD. This is harder to explain. But it could have been caused also by the European debt crisis, which began in 2009. Only in 2014 some of the effects of this crisis have been mitigated up to a certain point. The trend, afterward, was constantly growing.

A real prediction as far as the market evolution for Big Data is concerned, is hard to carry out, due to the pandemic. Paradoxically, this can be a cause for generating large volumes of data, but not necessarily from the industry. On the other hand,

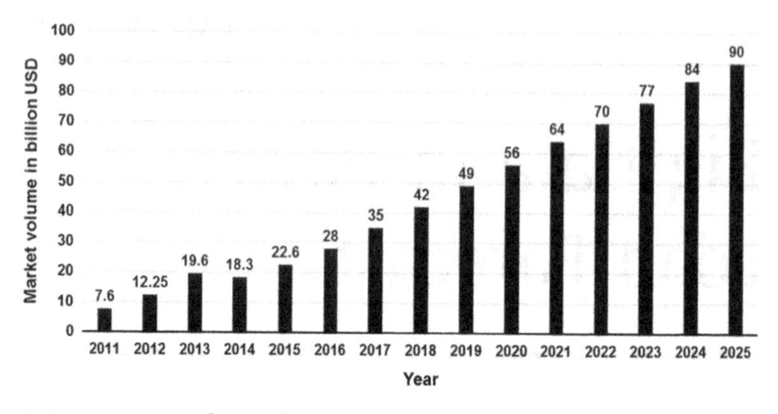

FIGURE 3.1 Market evolution for Big Data [1]

the consumption being low during this period, none of the industrial branches can keep the same pace as in the past. That means that the investments will lower but, in the end, the IT sector continues to remain strong. That is why, Figure 3.1 can give a rough idea about the market evolution for Big Data, even if this is not so more accurate as in the moment when the prediction has been made.

At the same time, in order to have a better knowledge about the impact of Big Data on the power engineering sector, with respect to other industrial branches, Figure 3.2 can be employed.

According to this study carried out in January 2019, the most important sector from the point of view of Big Data is Information and Communications Technology. This has been ranked I. Then, in the second class, are the financial and the trade sectors. That is obvious given the fact that the various transactions generate large volumes of data of all kinds. Possessing an impact graded 8 out of a maximum of 10, one has the category ranked III, in which the production (which is closely related to both previous fields), environmental protection, and healthcare can be found. In the middle of this ranking, one has the power industry which can be still considered an emergent field from the Big Data perspective. Its impact is assessed to 7.5. But being an

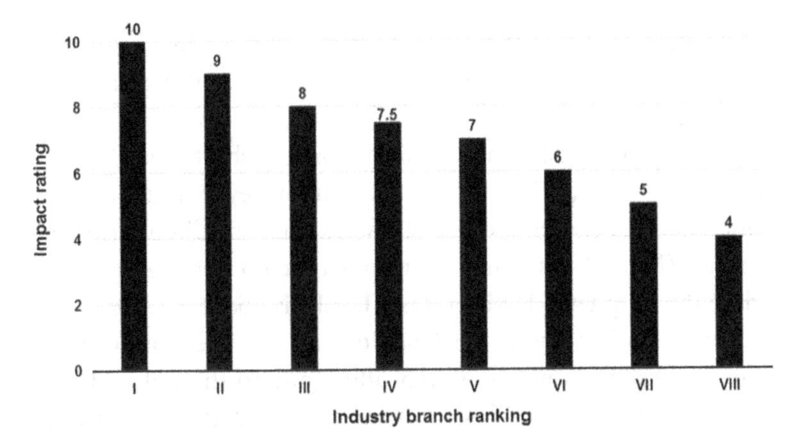

FIGURE 3.2 The impact of Big Data in power engineering [2]

important domain, generating enormous revenues, and because of the introduction of smart-meters, PMU, or SCADA Systems, its place within this ranking will change, perhaps, radically over the next years. On the other hand, the pandemic has altered decisively the consumption profiles. Many of the employees working from home, at present, consume more energy than in the past while the industry consumes less. Given this situation, over the next years, it is possible that new tariff regulations for electrical energy will be in place and these can be decided based on more data. Class V is represented by three industrial branches that possess the same impact of 7. These are the consumer electronics, the automotive, and the defence industry. The next in line is agriculture which is in class VI. Tourism and building form class VII, and, finally, the mining industry can be found in class VIII. Their impact rating is equal to 6, 5, and 4, respectively.

As in the other fields, also in power systems, the most important application of Big Data analytics is the load forecast. That is because it represents a very useful instrument when deciding the price of energy.

When it comes to Big Data processing in power systems, ML algorithms are always used. Among these, the Artificial Neural

Networks (ANNs) is the most encountered method for the very simple reason that they become more accurate the greater the training data set is.

Most of the load forecast is carried for very short term (VSTLF) and short term (STLF), at the distribution level. That means a time interval between 1 day and 3 weeks. The medium-term load forecast (MTLF) has a time horizon between 3 weeks and 3 years while the long-term load forecast (LTLF) is above 3 years.

On the other hand, the load forecast for the power transmission system is not done too often, due to the reduced data sets. This type of forecast will be described in detail within section 3.2.

Even the load forecast, at the distribution level, can be done with or without smart meters data. That is because, sometimes, the utility companies are very reluctant in making public this type of data, due to privacy concerns [3]. In a competitive retail market, consumption is influenced by the number of customers [3]. Thus, in Ref. [4], a two-stage LTLF is conducted that takes the consumer attrition into consideration. The first stage supposes the use of the linear regression, mentioned in section 2.2. Within the second stage, the consumer attrition is predicted through survival analysis. In the end, the result of both these predictions represented the load forecast. Even if this model is not trained with Big Data, is fully compatible with this concept. Usually, the volume of the training data in such cases is strictly correlated with the volume of exogenous data. Most of the time, in order to reduce the model complexity, only the temperature is considered. At the same time, another major issue is to choose exactly what kind of temperature parameters one should select when carrying out the load forecast. Related to this, is also the determination of the number and of the relevant meteorological stations [3]. As one shall see within the next section, the most efficient approach is to not use a unified weather index (such as, for instance temperature humidity) but only one. Thus, one solution is to aggregate these meteorological stations based on a Greedy algorithm [5] and then to rank them.

Other approaches suppose conducting the forecast for a substation within the grid or to prior process the data [3].

Another solution to tackle the problem of the weather parameters (for the day-ahead prediction) and their inclusion within the forecasting model is to use additive functions and to treat each moment of the day separately, as the consumption patterns change over 24 hours [6]. Thus, the forecasted load can be described using (3.1):

$$\hat{y}_t = c(t) + f(T1, T2) + D(t) + e_t \tag{3.1}$$

where:

$\hat{y}_t$-the forecasted for various time intervals (usually half an hour)

c(t)-a model that contains information with respect to the calendar

f(T1, T2)-a model that contains a temperature vector in location 1 and a temperature vector in location 2

D(t)-a model that contains information about the demand with respect to the time

e_t-an error vector.

The model *c(t)* is again additive and can be described through (3.2) [6]:

$$c(t) = d_t + w_t + s_t \tag{3.2}$$

d_t-takes a different value, depending on the day of the week

w_t-a variable that describes the holiday effect

s_t-is a function for a smoothing effect

For the temperature model, the arithmetic mean and the difference between the two temperatures are used.

The ANNs can be used also in the case of the load forecast without smart meter data. A very simple and straightforward method is to use a feed-forward ANN based on one layer of perceptrons [7]. The basic formula used in forecasting the load for this specific case can be seen in (3.3).

$$f(x, \omega) = \sum_{i=1}^{n} \omega_i \cdot \tanh\left(\sum_{j=1}^{m} \omega_{ij} \cdot x_j\right) + \omega_0 \qquad (3.3)$$

where:

n-the number of perceptrons

ω_i-the perceptron weights

m-the number of endogenous and exogenous variables of the model

ω_{ij}-the weights of endogenous and exogenous variables

x_j-the endogenous and exogenous variables of the model

Theoretically, the training, like practically in any other similar case, where one has to carry out a load forecast consists of minimizing the difference between the predicted values and the actual ones [7]. This difference can be considered a cost function. In order to minimize it, Levenberg-Marquardt, the conjugate gradient, or other similar methods can be employed.

In case there is no smart meter data at the disposal, one last approach for the load forecast is to make use of the node hierarchy within the grid. The method applied is represented by the Wavelet Neural Networks [8].

On the other hand, the smart meters can generate even TB of data as far as the distribution grids are concerned. That is why this type of equipment is completely compatible with the concept of Big Data. Using NILM combined with these data, one can obtain a very accurate forecast even of the individual appliances in the house. At the same time, the high granularity of the smart meter data improves the load forecast at aggregate levels [3].

In the remainder of section 3.1, various methods for the load forecast, based on smart meter data will be investigated.

At present, the forecasting methods based on aggregation can be or cannot be appropriate, depending on the situation. As in the case of the load forecast not based on smart meter data, also here the majority of the mathematical approaches make use of neural networks, training, and testing.

In 1992, a new prediction method called Support Vector Regression (SVR) has been developed. Its objective is to minimize the probability for the model generated from a certain training data set will commit errors on new examples [9]. Thus, the convex objective in (3.4) will be minimized, taking into account also the constraints (3.5) and (3.6).

$$f(x) = \frac{1}{2}\|w\|^2 + P\sum_{i=1}^{n} x_i + x_i^* \tag{3.4}$$

$$y_i - w^T\varphi(\overrightarrow{m}_i) - b \leq \varepsilon + x_i \tag{3.5}$$

$$w^T\varphi(\overrightarrow{m}_i) + b - y_i \leq \varepsilon + x_i^* \tag{3.6}$$

where:

w-regression weights

P-penalty factor

ε-desired error range for all points

x_i, x_i^*-variables that guarantee that at least a solution exists for all ε

$\varphi(\overrightarrow{m}_i)$-a mapping function of the input space to a higher dimensional feature space

The greatest advantage of this method is that there is a sole solution that minimizes the convex objective function [9]. Unfortunately, this is very dependent on the parameters P, ε as well as the parameters for the mapping function. In order to determine these parameters, one has at the disposal a plethora of algorithms like leave-one-out, grid search, and others [9]. But the greatest disadvantage of all is that SVR, given its complexity, is not appropriate to work with large data sets. That is why, in this case, one solution is data sampling. In this sense, any of the data sampling techniques described in the previous chapter can be employed.

Another method, similar to the one just described, is the Least Square Support Vector Machine. The differences are represented by the fact that the objective is based on least squares and the constraints are given in form of equality. In this sense, refer to (3.7) and (3.8) [9].

$$f(x) = \frac{1}{2}||w||^2 + P\sum_{i=1}^{n} x_i^2 \tag{3.7}$$

$$w^T \varphi(\overrightarrow{m}_i) + x_i + b = y_i \tag{3.8}$$

In this case, quadratic programming is not required to solve this problem. The algorithm solves a set of linear equations and thus that means that it performs faster than the previous one. Another important advantage is that this algorithm uses all its data to define the solution. Thus the scalability problem does not exist here and there is no need for any data sampling techniques.

Another interesting approach is the Hierarchical Mixture of Experts (HME). This is in fact an ANN that is trained to partition the input data coming from the smart meters across a set of experts. This type of approach is very useful when dealing with consumption data subtypes [9]. For example, subtypes dedicated to various seasons over the year like spring, summer, autumn, and winter. It is a known fact that consumption varies as a function of the season. Thus, a submodel emerges which is then assigned to an expert.

The HME network consists of two other types of ANNs: Gating and Expert. A possible approach in constructing this HME network can be as in (3.9) [9]:

$$F = \sum_i g_i \sum_{(j|i)} g_{(j|i)} \varphi_{ji}(\vec{x}) \tag{3.9}$$

where:

g_i-the Gating network output

$g_{(j|i)}$-the outputs from the lower Gating networks

$\varphi_{ji}(\vec{x})$-the output of an Expert network

The goal of the Gating network is to partition the input data over other Gating networks (if there are any) or among Expert networks. The activation function for the Gating network is the softmax function, as in (3.10):

$$\sigma_k = \frac{e^{x_k}}{\sum_{i=1}^{n} e^{x_i}} \tag{3.10}$$

where:

σ_k-normalized weight associated to the *k-th* subnetwork

n-the total number of subnetworks

x-the Gating network outputs

The scope of the Expert network is to specialize over the regions defined by the Gating network. This increases the efficiency and the accuracy of the forecast.

Another approach is represented by the Self Recurrent Wavelet Neural Network (SRWNN) [3,10]. In this case, the activation function consists of a wavelet function of the hidden neurons of a Feed Forward Neural Network (FFNN). Given the capability of adapting the wavelet shape to the training data set, this type of ANNs possesses a better generalization potential than the classical FFNNs. Hence, they are very appropriate when it comes to time series and load forecast [10]. At the same time, they are characterised also by the properties of the Recurrent

Neural Networks (RNNs) being able to store historical information about the wavelets. This is also due to the self-feedback loops. Another advantage of this kind of ANNs is that it is not necessitating a thorough parametrical analysis in order to achieve good results. These can be obtained based on the Morlet wavelet activation function for the neurons [10], as described in (3.11):

$$f(x) = e^{-0.5x^2} \cdot \cos(5x) \tag{3.11}$$

Given its characteristics, each current wavelet node is derived from its mother, according to (3.12) and (3.13):

$$f_{i,j}(n_{i,j}) = f\left(\frac{in_{i,j} - s_i}{sc_i}\right) \tag{3.12}$$

$$in_{i,j} = x_j + f_{i,j} \cdot z^{-1} \cdot p_{i,j} \tag{3.13}$$

where:

$f_{i,j}$-the scaled and shifted version of the Morlet mother wavelet

s_i-the shifting operator

sc_i-the scaling operator

$in_{i,j}$-the inputs of the wavelets

z^{-1}-the time delay

$p_{i,j}$-the weight of the feedback loop (the rate of information storage)

Taking into account what was mentioned previously, the term $f_{i,j} \cdot z^{-1}$ represents a memory term dedicated to storing the historical information about the wavelets. On the other hand, the rate of information storage, $p_{i,j}$, represents the most important difference between a regular Wavelet Neural Network (WNN) and an SRWNN. The SRWNN behaves identically to a WNN when all $p_{i,j} = 0$ (this, in spite of the fact that initially all $p_{i,j} = 0$) [10].

In the end, the predicted value $\hat{y}$ is given by (3.14):

$$\hat{y} = \sum_{i=1}^{n} w_i \cdot f_i + \sum_{j=1}^{m} v_j \cdot x_j + b \quad (3.14)$$

where:

n-the neuron number
w_i-the weight between the neuron i and the output node
m-the dimension of the wavelet function
v_j-the weight between the input j and the output node
b-the bias of the output node

Other advantages presented by this kind of ANNs are the capability of sensing not only the cyclical behaviors but also the trends within the training dataset. At the same time, not being dependent on the fine-tuning of the parameters like the SVR method, described earlier, it is compatible with the concept of Big Data. Another feature which makes the SRWNN even more attractive from this point of view is its capacity for storing historical data.

When it comes to load forecasting at the household level, sparsity is an important condition. This characteristic is again consistent with the concept of Big Data. Hence, the forecast can be carried out also taking into account the consumption of other houses within the neighborhood and at the same time the sparsity of the interconnection between houses [11]. Exploiting further this assumption, it is considered that the forecast for the target house is biased only by few other households.

Another important aspect that has to be taken into account is the fact that at the household level, the volatility of the load and its stochastic character are more obvious than at a higher level of consumer aggregation [12]. This is practically another reason for which the sparse coding can prove an efficient technique for load forecasting at the individual household level. But even in this case, in order to attain a better accuracy, this forecast should normally be carried out over shorter periods of time. In order to model the consumption patterns, the basic idea is to learn a dictionary of a certain number of vectors to represent frames with the same hour duration as the vector dimension. These hours are sampled from the loads. In this sense, one can use either basic sparse or group sparse [12]. In the first situation, the reconstruction error is characterized by a square loss. This model introduces a penalty factor for sparsity as well and non-negative constraints for the dictionary elements and the decoding coefficients [12]. The group sparse, on the other hand, replaces the sparsity penalty factor with a group of penalty factors which force the frames resulting from the same meter to use the same dictionary elements in the signal decomposition [12]. This type of approach is also consistent with the Big Data concept. According to [12] data resulting from 5000 households has been used in order to achieve an improvement in accuracy of 10%. In reality, there is no limitation in this sense. The greater the data set is, the better the forecast precision will be.

Another approach in the sparsity use within the field of load forecast at the household level, is to adaptively explore this sparsity in the training data set [13]. Thus, the sparsity is captured by Least Absolute Shrinkage and Selection (LASSO) estimator. Practically, the goal of LASSO is to select relevant lag order within the training data set for avoiding model overfitting. This is not very easy to do and because LASSO is a convex relaxation to this problem, but its use is very suited. The mechanism used by this method consists of shrinking the estimated parameter towards 0 [13] for avoiding the model overfitting and to select the relevant regressors. It turns out that when LASSO is applied to the autoregressive model, this method attains the best accuracy among all the linear models [13].

Indirectly, the Big Data concept is very useful also when considering the noise reduction of the smart meter data. That is because a highly employed approach is to aggregate the data resulting from individual households [3]. Unfortunately, here the risk of accidentally removing the trend in the data is important. Thus to avoid a situation of this type, clustering can be a solution.

But before the clustering is carried out, the load forecasting can be formulated as a functional problem [14]. For the sake of simplicity, the most direct solution is not to take into consideration any problems related to the calendar (transitions from a working day to a free day or vice versa). Thus, if the electricity demand represents a continuous-time process as in (3.15) [14]:

$$x = x(t) \tag{3.15}$$

where:

x-the electricity demand to be predicted

t-the time interval

one needs to predict the variation of x over the time frame $[t; t+d]$. In this sense, one solution is to divide the interval $[0; t]$ in subintervals [14] of the form $[i{\cdot}d; (i+1)\cdot d]$, $i=0, \ldots, j-1$ and $j = \frac{t}{d}$.

Conversely, as far as the load forecast itself is regarded, one can consider the demand in each moment as a sample corresponding to the time moment t, according to (3.16) [14]:

$$s_n(t) = x\cdot[t + (n-1)\cdot d];\ t \in [0; d),\quad n \in N \tag{3.16}$$

So, if the initial observation of the measured time series is denoted as in (3.17), then the general form of the forecast can be described based on (3.18).

$$s_{ni} = \{s_1(t_1),\ s_2(t_2),\ s_3(t_3),\ \ldots,\ s_M(t_M)\} \tag{3.17}$$

$$\hat{s}_n = \{\hat{s}_1(t_1),\ \hat{s}_2(t_2),\ \hat{s}_3(t_3),\ \ldots,\ \hat{s}_M(t_M)\} \tag{3.18}$$

where:

s_{ni}-the measured data
M-the total number of samples
$\hat{s}_n$-the predicted load

An efficient forecast method when transforming the time series into a functional problem can be also SRWNN but is not limited only to this. Otherwise, easier to implement can be a regression method. In any case, this functional problem formulation is fully compatible with the Big Data concept, as it is not influenced by the data volume.

Despite the fact that the clustering increases the forecast accuracy, it increases also the computation time. So, if time is a constraint, a trade-off between the two must be determined.

Even the clustering techniques differ and there is no definite answer which is the best, as these techniques are always influenced by the specific situation they solve.

A novel approach to clustering will be described in the sequel. The goal of this approach is to forecast the energy consumption for the next 30 minutes either at the level of an individual household or for a group [15]. Hence, after the removal of bad data from the training set, certain consumption patterns emerge from the series. Within the next step, these are clusterized and thus various classes of pattern types come into being [15]. Based on some exogenous variables, depending on the day, a second stage clusterization is conducted for these classes. At the next step, a forecast model for the new clusters is generated. Basically, these models are based on very short historical load data and are made at a minute [15]. To put it short, for this kind of approach, Big Data can be employed only within the initial stage, before the bad data removal. Then various prediction algorithms can be employed, even SVR, since after clusterization the data volume decreases. The novelty of this approach consists more in data organization and preparation which are very thorough but necessitate important computation resources. So, one can say, that this method is appropriate only when the time does not represent an essential constraint.

Another type of approach is the one that makes the correlation between energy consumption and temperature based on the adoption of electrical air conditioning (AC) units [3,16]. But this research direction presents certain limitations. That is because in some northern countries, like for instance, the Netherlands, Germany, or even France such units are not used extensively. According to recent statistics conducted in 2018, only one-third of the population at the global level own such an installation. It is true that in countries like the USA or Japan, 90% of households possess AC. At the same time, mention should be made of the fact that given the current pandemic situation which has radically altered the consumption patterns, the adoption of such installations is susceptible to decrease in the future. So, the correlation between electrical energy consumption and temperature must not be studied strictly from the perspective of the AC units.

A very novel approach for the load forecast consists of taking into account the so-called recency effect [16]. Similar to the psychological studies which came to the conclusion that humans remember most of the time only the recent events, also the power grid behavior tends to be influenced, above all, by recent temperatures.

In the end, this recency effect can be coupled with all types of algorithms dedicated to the load forecast. Up to now, the Tao Vanilla Benchmark Model which is one of the best forecasting methods applies it together with the Multiple Linear Regression Model given its simplicity as far as the implementation is concerned. This type of regression can be characterized using (3.19) [16]:

$$\hat{y}(t) = c_0 + c_1 \cdot trend(t) + c_2 \cdot m(t) + c_3 \cdot w(t) + c_4 \cdot h(t) + c_5 \cdot w(t) \cdot h(t) + T(t) \quad (3.19)$$

where:

$\hat{y}(t)$-the load forecast corresponding to the time moment t

c_i-the coefficients estimated through the Least Squares Method

$m(t)$-the month of the year

$w(t)$-the week of the month

$h(t)$-hour of the day

$T(t)$-temperature corresponding to the time moment t

Given this model, one approach is to use only historical temperatures and calendar information (regarding working days, free days, or holidays), without any electrical loads [16]. The lack of historical load data is compensated by the recency effect applied to the temperature. But even so, this kind of approach has some drawbacks [16]: it under-forecasts the troughs over consecutive days; it over-forecasts the summer peaks and under-forecasts the winter peaks over consecutive days; the forecast leads or lags the actual load over consecutive hours. Another disadvantage is that coupling the recency with the Multiple Linear Regression Model results in a forecast for which, in order to be optimized, more than a thousand parameters have to be included. An explanation can be the insufficient data volume. So, also here the Big Data concept comes to play an important role as far as the improvement of the prediction accuracy is concerned. As it will be shown in section 3.2 of this work, the CNNs take into account this characteristic right from the start, attaining a very good accuracy even when not trained with exabytes or petabytes of data.

Another important aspect is how to include in the forecasting model also other factors related to weather, besides the temperature.

As has already been mentioned within this section, one proposal was to employ a unified temperature–humidity index. But this turned out to be not a viable solution when it comes to load forecast [17].

The forecast is also here carried out through the Multiple Linear Regression Model.

The conclusion was that it is better to separate the temperature, relative humidity, and their higher order terms. This together with the corresponding coefficients in (3.19) being estimated based on the training data, led to better accuracy of the forecast.

Another type of approach was to observe the interaction between the load and the wind speed. It is known that when it is cold and, at the same time, the weather is windy, one tends to feel the temperature lower than this actually is. In this case, the load tends to rise. Hence, a so-called wind chill index (WCI) has been introduced in USA by the National Weather Service (NWS) [18]. Its formula can be seen in (3.20):

$$WCI = \begin{cases} 35.74 + 0.6125 \cdot T(t) - 35.75 \cdot WS_t^{0.16} + 0.4275 \cdot T(t) \\ \quad \cdot WS_t^{0.16},\ T(t) < 10°C \\ \quad T(t),\ otherwise \end{cases} \tag{3.20}$$

where:

$T(t)$-the air temperature in °C

$WS_t^{0.16}$-the wind speed at the time moment t

In (3.20), to the condition that $T(t) < 10°C$, one can add also that $WS_t > 4.8$ km/h.

Practically, the index presented in (3.20) must be extended and correlated with the load. The easiest approach is to introduce an additive model [18].

For Canada, this formula is slightly different as far as the coefficients are concerned. The experiments have shown that instead of including just the WCI into the forecast, it is better to separate the wind speed from the temperature and their variants in order to obtain a good accuracy of the prediction [3,18].

As mentioned previously, another problem to be tackled, when it comes to load forecast, is the proper selection of the weather stations. That is because, especially when the forecast is carried out at the transmission level, the operator can cover more than one geographical area [3]. Thus, a straightforward approach is to decide how many weather stations have to be included and which of them are relevant [19].

Another aspect that has to be taken into account is the fact that currently, the distributed energy resources (DER) present a high penetration in the power grid. From this point of view, solar generation can be an attractive solution. But this, of course, introduces other forecasting problems related above all to weather and also to energy prices. Wind generation poses similar issues. From here, the necessity of load forecast for a certain type of generation emerges [3].

This kind of problem could be solved by clustering the various types of consumers, connected to a specific generation type, use a certain method for each of the clusters, and, in the end, obtaining the forecast based on the aggregation of the individual predictions.

Another important issue related to all forecasting methods is the measurement of their performance. In the majority of the cases, this is done by using the Mean Absolute Percentage Error (MAPE) [3].

In the case of forecasting aggregation, some studies have shown that MAPE decreases rapidly when the number of consumers, for which the forecast is carried out, increases (the total number here should be less than 100000) [3]. If the customer number is greater than 100000, an increase in the number of customers will cause a little decrease in the MAPE. That is why also, from this point of view, the concept of Big Data can play an important role. But here, the discussion is much more complicated because the decrease depends also on the volume of data coming from each customer. So, the number of clients is not the only relevant parameter.

As one shall see, within the next section, the performance evaluation of the CNN is based also on MAPE. As a rule, if this parameter is lower than 10, one considers the forecast as highly accurate [20]. If it is between 10 and 20, then the accuracy is good; between 20 and 50, the performance is reasonable and, finally, if MAPE is greater than 50 then the prediction is considered inaccurate.

An alternative to MAPE could be the resistant MAPE (the r-MAPE) resulted from the calculation of the Huber M-estimator [20]. This version of the parameter behaves better when it comes to outliers.

There is also a geometric variant that treats better the intermittent character of the individual load profiles [21]. It is called MAAPE and stands for Mean Arctangent Absolute Percentage Error. This geometric variant considers the tangent of the angle defined as the ratio between the absolute error and the actual value.

Combining different types of forecasting methods can improve the forecasting accuracy even when this carried out using Big Data. Thus, an interesting approach that can be used together with the classical methods based on ANNs is probabilistic forecasting. This is especially consistent with the concept of Big Data since based on these probabilistic methods, many scenarios as far as the input temperature data can be obtained. For instance, the utilization of the historical temperature data with the dates fixed [22] or even shifting these data by a few days with respect to the previous year [23] in order to obtain a better variety of the input data and, at the same time, in order to try to better assess the uncertainty of the prediction even in the case of reliable methods like the CNNs trained with Big Data. It turns out that the shifting method is better when this shifting takes place within a certain range [23].

Another method dedicated also to the probabilistic load forecast is based on a hybrid model between an ANN and linear regression. The resulting method is called Quantile Regression Neural Network (QRNN) [24]. The advantages of this approach consist of the fact that both the uncertainties related to the temperature variation and those related to the load variation are overcome. The results are generated in form of quantiles. The normalized hourly variables (like, for instance, the temperature) represent the training features while the corresponding hourly loads are the training labels. The training process ends when the loss function no longer decreases [24], meaning that the difference between the predicted value and the actual one cannot be improved anymore.

The load forecast, especially at the distribution level, can be also directly influenced by the reliability indices of the system, like, for example, the System Average Interruption Duration Index (SAIDI). This is defined as in (3.21).

$$SAIDI = \frac{Sum\ \ of\ \ all\ \ Customer\ \ Interruption\ \ Durations}{Total\ \ number\ \ of\ \ Customers\ \ Served} \tag{3.21}$$

The variable to be forecasted can also be the SAIDI [25]. The numerator in (3.21) represents the outage period measured in minutes over a certain reference interval which is usually 1 year. But, this type of parameter definition is not constrained only to this. There are also cases in which SAIDI can be calculated monthly or even daily [25]. These are used typically for improved system reliability analysis. Obviously, the greater this indicator is, the worse the customer experience will be.

But, in the end, this kind of prediction is only indirectly related to the actual load forecast.

As it has already been seen within this section, the majority of the load forecast procedures are applied at the household level. That is because the data coming usually from the smart meters is easier to obtain, diverse, and already in large volume. This means automatically also a better sampling rate which in turn allows a more accurate prediction. But when conducting the load forecast at the household level there are also some disadvantages. One of them is the short or very short time interval over which the prediction is carried out. Why is that? The answer lies in the fact normally the data volume coming from only one smart meter is in the range of GB only for one day (depending on the setting up of the smart meter parameters, it can be even more). Carrying out a forecast for one year becomes very difficult because, in order to do this, one has to use at least one year of historical data to have a minimum of accuracy for the prediction. Thus, the potential of Big Data becomes enormous for situations of this kind. Not only from the point of view of the volume storage but also from the point of view of the data variety. For instance, in very few papers, when it comes to load forecast, this is treated in relation to calendar problems (alternation of working days and free days or even the alternation of holidays with longer working periods over the year). Big Data has the potential of overcoming this obstacle too, delivering forecasts that take into account the changes in the consumption profiles induced by these aspects related to the calendar.

As stated previously, a novel method that yields interesting results and is fully compatible with the concept of Big Data is based on CNN. One of the most important advantages, in this case, is that it offers the flexibility to easily include various parameters related either to seasonality or to calendar issues. To be more specific, the transitions between the free and working days or vice versa.

As it will be seen in section 3.2, the load forecast will be carried out at transmission-level, for 1 year.

3.2 BIG DATA IN POWER LOAD FORECAST AT POWER TRANSMISSION LEVEL

Initially, the CNNs which will serve as a method for carrying out the load forecast have been initially developed for image recognition. There are numerous applications in which the image used as input for the CNN is then classified at the output. For example, one of the most encountered examples is when at the input a scanned handwritten figure is used and then this has to be "recognized" by the CNN and correctly classified as "0", "1", …, or "9."

The method based on CNN has been chosen because this type of ANN offers important capabilities of deep learning, has a good accuracy as far as the forecast is concerned, offers a good calculation speed and that is why it is implicitly compatible with the concept of Big Data.

But there is also another, more specific reason. This is represented by the fact that the two consecutive pixels within an image possess the same correlation as two consecutive load values within a time series. This makes CNN an optimum method for the forecast in general.

In the sequel, the architecture of a CNN dedicated to image processing will be described, after which the differences between this and a CNN dedicated to load forecast will be investigated.

Let's consider that at the input of the neural network, one has an image that is characterized by a certain width, W, and height, H. If one has n neurons in the first hidden layer of the CNN, each of these neurons will be connected to all pixels in the image, resulting thus $W \times H$ connections [26]. If the image is 64 x 64 pixels this will be in the end characterized by a unidimensional vector of 4096 values. If the first hidden layer in the CNN has, for instance, 8000 neurons and each neuron is connected to every pixel in the image, one shall have 4096 x 8000 = 32768000 of connections and, implicitly, of parameters. And this number is not even close to the case of an HD image. So the computational burden increases and a solution has to be found in order to mitigate it. One obvious way to do it is to reduce the number of neurons in the hidden layer. Bu this will affect negatively the performance of the CNN. Thus the number of neurons has to be maintained rather high [26]. Another possible solution is to group these neurons into blocks taking into account the fact that neighboring pixels are highly correlated and that the pixels found in distant areas of the image are not. In this situation, in order to gather the information from consecutive pixels, one uses filters (kernels) of usually 5 x 5. This kind of filter is dedicated to determining the content in an image. If one wants to determine the edges of the image, the dimension has to be 3 x 3. Using this type of approach, the number of parameters is dramatically reduced. On the other hand, each neuron in the block possesses different weights. If one supposes that all the neurons in the same block have the same weight, the number of parameters and thus the computational burden can be reduced even further. This procedure is called weight sharing. Applying these two techniques (block partition and weight sharing) simultaneously, one can obtain a parameter reduction of 99% compared to the fully connected layer [26].

The output of each neuron, before the weight sharing can be described based on (3.22) [26].

$$f^l_{a,b} = \sum_{i=1}^{5} \sum_{j=1}^{5} img\,(a + i,\, b + j)\, w^l_{i,j}) \tag{3.22}$$

where:

$f^l_{a,b}$-the output of the neuron *(a,b)* in block *l*

$w^l_{i,j}$-the weight of the neurons in block *l*

In equation (3.22) each of the sums varies between 1 and 5 because each neuron is connected to a 5 x 5 region of the image. This region is called *receptive field*. Another important aspect is that the output of each block has the same size as the block itself. This is not necessarily 5 x 5 [26]. It can have also another value.

The equation (3.22) practically gives the name of this type of ANN since it is analogous to convolving a 5 x 5 kernel with the input image [26]. This implies that the convolution layer output is generated by convolving each filter on the input image. Thus the result is represented by a series of images in which each of these is obtained through a kernel. In other words, in this case, the number of images is equal to the number of filters [26]. The next step is to obtain a feature map of the images. To this goal, an activation function will be applied to them. In this way, if an image is characterized by a certain width and height ($W \; x \; H$), and the convolution layer consists of L filters, of size $P \; x \; Q$, the result of the convolution layer will be represented by L images of size

$W - P + 1 \; x \;\; H - Q + 1$. As already mentioned, each image is generated by convolving the corresponding kernel with the input image.

Usually, a CNN consists of two to three convolutional layers. The practice tends to confirm that the accuracy of the classification or of the forecast tends to worsen when the number of convolutional layers increases.

Basically, for a CNN, a kernel represents a 2-dimensional array that is applied to a grayscale image. If the input is represented by an RGB image, the convolution filter is again a 2-dimensional array that will be separately applied to each channel [26]. The most important aspect here is that the result of the convolution for a multichannel input (such as an RGB image) has always a single channel. At the same time, one can consider the RGB image as a three-dimensional array in which the first two dimensions contain the spatial coordinate of the pixels while the third indicates the channel [26]. For instance, an HD image of 1280 x 720 pixels will be stored in a 720 x 1280 x 3 array.

Generally, for an image with multiple channels, the filter has to be three-dimensional and the third dimension is always equal to the number of the input channels [26]. Given this property, the CNNs have also multiple convolution layers. Hence, each filter will be convolved with the input image. As mentioned previously, the number of images will be equal to the number of filters and, further, the pixels of each image will be pooled in order to extract the information more efficiently. The practice shows that when the pooling is used, the accuracy of the classification or of the forecast tends to diminish but the computation speed is much improved. So, the decision about using the pooling layer depends on the nature of the application and, finally, on a compromise between the two objectives. Coming back to the convolution, some filters in the first layer act as low-pass filters (smoothing filters) while others as high-pass filters (dedicated to the edge detection of the image) [26]. In reality, a CNN has the capability of learning to adjust the weights of the filters [26] during the training. The end result of this is that the classes can be linearly separated within the fully connected layer. The training algorithm consists of forwarding propagation toward the convolutional layer (and eventually the pooling layer) of the information that characterizes the image and backward propagation. The latest type of propagation has the role of updating the weights of the filters and takes place based on the gradient of a loss function. Basically, within the current convolution layer, the gradient of the loss function is equal to the convolution between the vector of gradients in the current layer and the reverse of the convolution filters.

Another important element that has to be taken into consideration when it comes to convolution is *stride*. This, practically, controls how the filter convolves. In other words, it fixes the number of shifting positions of the filter. In order to obtain the best accuracy, it is usually set to 1. In this way, all the areas of the image are covered efficiently. But if one wants to compute the convolution of alternate pixels, then a larger stride has to be chosen. The most important aspect related to stride is that this must be chosen in such a way that the output of the convolution is an integer value and not a fractional one. So, here, also the filter dimension is important. In the case of a fractional value, the image has to be cropped [26].

Related to the stride, is also the pooling which was introduced previously within this section. Its goal is to reduce the dimension of the feature maps. That is why sometimes it is called also *downsampling*. Given this, the factor used to carry out the downsampling is the stride. So, if one chooses, for example, the stride to be equal to 3, this means that beginning with the first, every third pixel in the image will be selected. In this way, the image and implicitly the information characterizing it, become 3 times smaller than in the case one would have chosen the regular value 1 for the stride. But using pooling and striding simultaneously, without any other rule can lead to a loss of relevant information. That is why a rule which is usually introduced in order to consider the information between alternate pixels is the *maxpooling*. Another important aspect that has to be considered is that overlapping of the pooling regions has to be avoided. If the stride equals the downsampling size, no overlapping arises.

As stated previously, the aim of the pooling is to reduce the dimension of the feature maps; this can lead instead to a lack in accuracy in classification or prediction, depending on the operation for which the CNN is used. Due to this, sometimes, it is considered that the pooling layer can be replaced by a convolution layer with a stride equal 2 [26]. If the end result of the convolution is again an image, one cannot avoid the pooling. Instead if one uses the CNN as a method of prediction, and especially if the input is represented by time series, then one can renounce to the pooling layer. This will be demonstrated later within this section taking as an example the load forecast in the Romanian power transmission system.

At the same time, there is also another possibility to consider the information between alternate pixels, besides the maxpooling. This is the average pooling and it has been shown that this delivers better results than the previous method since it takes into account more than one value in the region. The result of the pooling, in this case, is represented by the arithmetic average of values considered. Finally, there is also a stochastic pooling in which one parameter within a region is chosen based on a certain statistical distribution.

Coming to the problem of the load forecast where the input is represented by time series, it is important to know that the CNN behaves practically the same way as in the case where the input is represented by an image. Basically, the main idea is to extract the future consumption trend from the consecutive values in the time series that serves as input. This happens based on the convolution operation as in (3.23) [27]:

$$O(i) = (In * f)(i) = \sum_{m} In(i + m) \cdot f(m) \tag{3.23}$$

where:

$O(i)$-output of the convolution operation

$In(i)$-the unidimensional input

$f(m)$-the unidimensional filter

$O(i)$ can be considered also a feature map. The feature map [27] is thus obtained for our specific case by convolving always the same filter with the whole time series. The feature determined in this way will be learned, not taking into consideration its position within the time series.

Given the fact that a unidimensional time series is used as input, the chosen filter is also unidimensional. Generally, if one would have had an image instead of a time series one should have used a two-dimensional filter. Both types of filters present very good results in terms of computation burden. Another important aspect is that during the training process which in this case has been carried out through Adaptative Movement Estimation (ADAM) only the filter weights are updated [27]. This makes the CNN faster and more efficient with respect to memory. That is why this kind of ANN represents an obvious solution for load forecasting problems. The choice of a unidimensional filter, in this case, makes the CNN perform even better. The general architecture of the used CNN is presented in Figure 3.3.

As mentioned previously, the training is done with ADAM [28]. This algorithm combines the advantages of

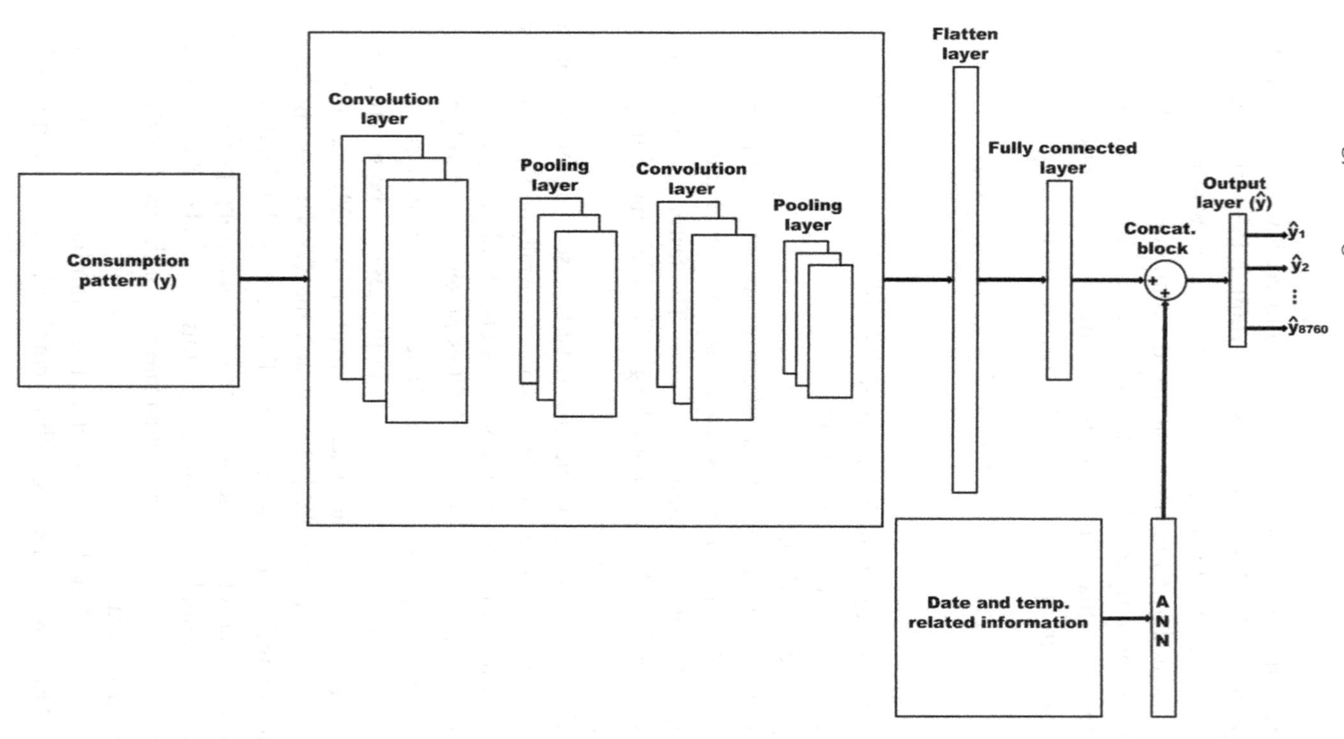

FIGURE 3.3 The architecture of the proposed CNN

AdaGrad and RMSProp [28]. Among the ADAM advantages, one can mention:

- the magnitudes of the parameter updates do not depend on the rescaling of the gradient
- there is no required stationary objective; the algorithm performs well with sparse gradients and, at the same time, conducts a form of size annealing.

Practically, these are the principles based on which the weights update takes place, an update which, in the end, can be regarded as a step-by-step procedure.

The algorithm updates the exponential moving averages of the gradient of the loss function, m_t, [28] as well as the squared gradient, v_t, at which the hyperparameters β_1 and $\beta_2 \in [0; 1)$ control the exponential decay rates of the moving averages. There are estimates of the uncentered variance and of the mean of the gradient.

The step size has the following upper bounds [28]:

$$|\Delta_t| \leq \alpha\Delta\frac{1-\beta_1}{\sqrt{1-\beta_2}} \text{if } 1-\beta_1 > \sqrt{1-\beta_2} \tag{3.24}$$

$$|\Delta_t| \leq \alpha \text{ if } 1-\beta_1 \leq \sqrt{1-\beta_2} \tag{3.25}$$

The size of the steps in order to mitigate the loss function is bounded by α. This is equivalent to the creation of a trust-region for the current value of the parameter. Outside this region, the estimate of the gradient does not possess enough information [28]. From this, it results that because the trust region is already known, the value of α can be determined. This is valid for the majority of the ML algorithms. Because α represents the upper bound of the step number mentioned above, its value can be determined iteratively.

The convergence analysis is carried out based on a regret function. This is defined as the difference between a prediction and the best-fixed point parameter, as in (3.26):

$$R(N) = \sum_{t=1}^{N} [f_t(a_t) - f_t(a*)] \tag{3.26}$$

where:

$f_t(a_t)$-the prediction

$f_t(a*)$-the best-fixed point parameter

As stated previously, there are also some methods that are related to ADAM, like the natural Newton method, RMSProp, AdaGrad, vSGD, or AdaDelta. All these methods share a common point: they are setting step sizes through the estimation of the loss function from the first-order information [28]. At the same time, there are also methods for the evaluation of the loss function that present linear memory requirements with respect to the minibatch partitions of the dataset. If one uses a GPU to train the CNN, this type of method is not recommended, because the GPU is memory constrained [28]. In this category, one has the Sum of Functions Optimizer (SFO) which, in reality, is based on the Newton algorithm and, at the same time, uses minibatches.

On the other hand, ADAM uses a principle that adapts itself to the data volume and type.

Usually, the most encountered alternatives to ADAM, are AdaGrad and RMSProp.

AdaGrad works very well for sparse gradients. It practically represents a version of ADAM in which $\beta_1 = 0$ and $1 - \beta_2$ is almost 0. Also, α will be replaced, in this situation, by an annealing expression of the form $\alpha \cdot t^{-1/2}$ [28]. In this case, when the bias correction is not considered, the correlation between AdaGrad and ADAM is not valid anymore.

As far as the RMSProp is concerned, this is even more similar to ADAM, especially the RMSProp with momentum. But there are also important differences [28]: the ADAM updates of the weights are estimated using a moving average of the first and second moment of the gradient while the RMSProp generates this update using a rescale of the gradient. At the same time, RMSProp does not possess a bias correction either. This will lead in the majority of cases to very large step sizes and most probably to divergence [28]. This specific situation appears when $1 - \beta_2$ is almost 0 (as in the case of the sparse gradients).

Applying ADAM to the CNN training shows promising results. What is characteristic of CNN is that the weight sharing generates different gradients in the various layers of the network. In spite of this, ADAM converges much faster than the other methods [28]. At the same time, it adapts the value of the learning rate for the layers of the CNN, instead of manually choosing them, like in the case of Stochastic Gradient Descent (SGD).

The load forecast studied in the sequel is based on a CNN training using three years: 2016, 2017, and 2018, with hourly loads. The load forecast has been carried out for 2019. This type of training is compatible with the situation in which the values used in this operation result from the various sampling algorithms employed in Big Data processing. The data has been taken from the website of the Romanian energy transmission operator.

The solution proposed here is very similar to the one in [27]. Thus the forecast is based on a single exogenous input, this being the temperature. The average values of the temperature have been used, based on the data obtained from the underground website. At the same time, the calendar-related information has been considered (like, for example, the day of the week or the season).

An important aspect, in this case, is that the exogenous information is introduced in the model by using another ANN, whose output is concatenated with the output of the CNN fully connected layer.

Again as in Ref. [27], in order to describe the day in week 7, binary inputs together with one-hot encoding are used (1000000 for Monday, 0100000 for Tuesday, and so on). In the case of the seasons, the methodology is identical. Hence, one employs 1000 for spring, 0100 for summer, etc. From the point of view of the holidays which can present special consumption patterns, the binary code is once more utilized in order to ascertain if a certain day is a holiday or not.

Regarding the performance evaluation, the present model is assessed based on Mean Absolute Percentage Error (MAPE), Mean Absolute Error (MAE), and Root Mean Squared Error (RMSE). Their formulas are described in (3.27), (3.28), and (3.29), respectively.

$$MAPE = \frac{1}{n} \sum_{1}^{n} \left| \frac{y - \hat{y}}{y} \right| \cdot 100 \tag{3.27}$$

$$MAE = \frac{1}{n} \sum_{1}^{n} |y - \hat{y}| \tag{3.28}$$

$$RMSE = \sqrt{\frac{1}{n} \sum_{1}^{n} (y - \hat{y})^2} \tag{3.29}$$

where:

y- the actual value

$\hat{y}$-the forecasted value

The CNN has been implemented in Python. The training lasted approximately 60 minutes. The computer is equipped with an Intel i5-8300-H 2.3 GHz processor and an Nvidia Geforce GTX 1050Ti 4Gb video card. In this case, both processors have been used and thus the computation time was reduced significantly. Based on this approach, the future CNNs that use Big Data for learning must concentrate on GPU clustering [27] in order to be more efficient.

At the same time, as far as the architecture itself is concerned, there is not any rule of thumb based on which one can exactly determine the number of convolutional layers and of the pooling layers or the number of kernels and their dimension. That is why 14 architectures have been investigated (with or without pooling layers) and these are presented in Table 3.1.

The interpretation of Table 3.1 is as follows. One takes, for example, architecture number 11. This is characterized by two convolutional layers. For the first layer, one will have 15 filters of dimension 5 and, for the second layer, one will have five filters of dimension 3. At the same time, the two pooling layers have a dimension of 2. On the other hand, architecture number 10 has again two convolutional layers but no pooling. Architecture number 5 possesses a single convolution layer with five filters of dimension 3 and a pooling layer of dimension 2.

TABLE 3.1 The Investigated CNN Architectures

Arh. No.	Conv. Layers	[No. of Filters, Kernel Size]	Pooling Size
1	1	[(5,17)]	[2]
2	1	[(5,17)]	-
3	1	[(5,9)]	[2]
4	1	[(5,9)]	-
5	1	[(5,3)]	[2]
6	1	[(5,3)]	-
7	2	[(5,17),(5,9)]	[2,2]
8	2	[(5,17),(5,9)]	-
9	2	[(5,9),(5,5)]	[2,2]
10	2	[(5,9),(5,5)]	-
11	2	[(15,5),(5,3)]	[2,2]
12	2	[(15,5),(5,3)]	-
13	2	[(15,9),(5,5)]	[2,2]
14	2	[(15,9),(5,5)]	-

In order to see which of these configurations is the best, one has to take a look at their MAPE, MAE, and RMSE performance indicators. To do this, one can observe Figures 3.4, 3.5, and 3.6.

As one can see in all these figures, the best architecture is the one in number 12 since this has the smallest MAPE, MAE, and RMSE. The values for them are 1.655%, 114.81 MWh, and 161.136 MWh respectively.

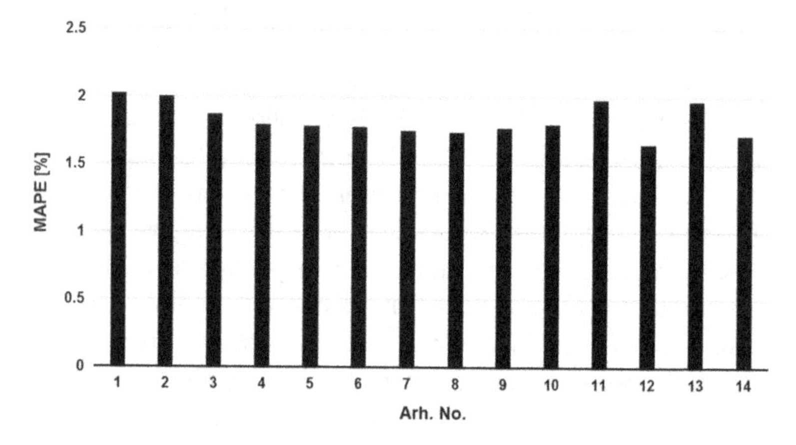

FIGURE 3.4 The MAPE performance indicator

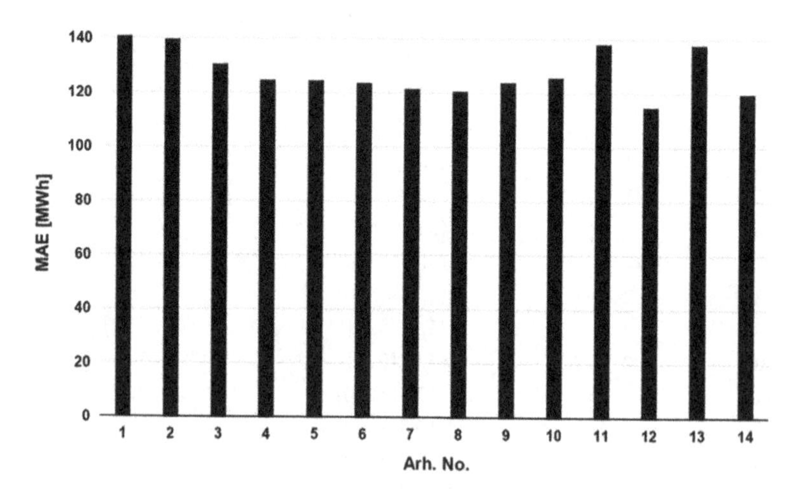

FIGURE 3.5 The MAE performance indicator

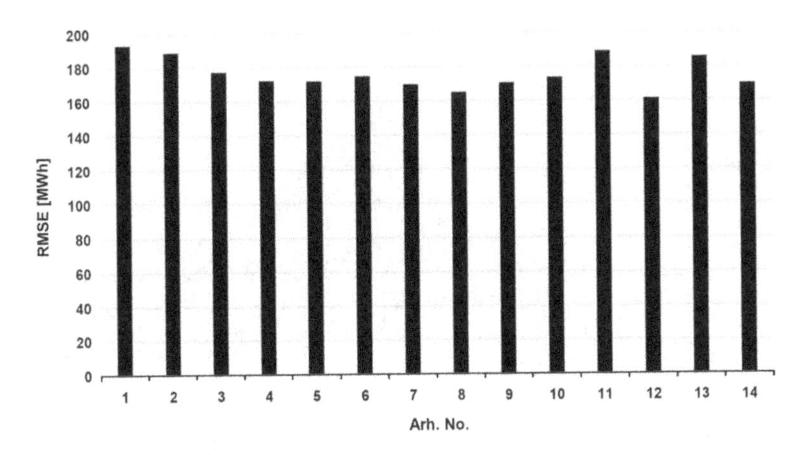

FIGURE 3.6 The RMSE performance indicator

The second best is number 14. The values for MAPE, MAE, and RMSE, in this case, are 1.72%, 119.7 MWh, and 169.7 MWh.

As one can see, none of the configurations possess a pooling layer. This type of layer is always equivalent to a loss in features and hence in forecast accuracy, given the reduction in the size of the convolutional layer output.

In the sequel, one shall present the results of the load forecast for these two configurations. These results can be observed in Figures 3.7, 3.8, 3.9, and 3.10.

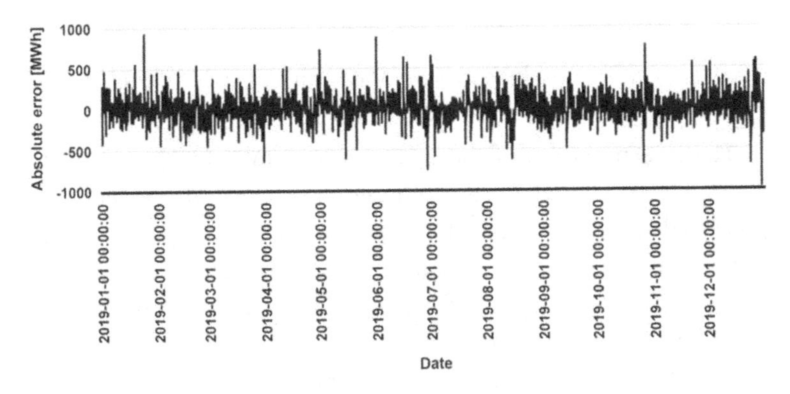

FIGURE 3.7 The absolute error in MWh for the architecture 12

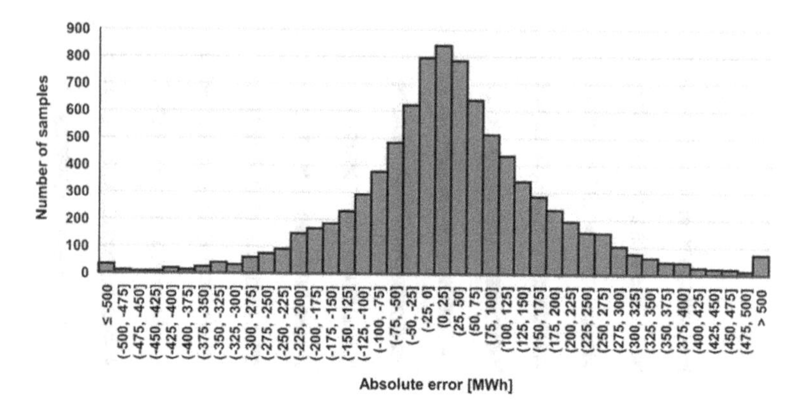

FIGURE 3.8 The error distribution for the architecture 12

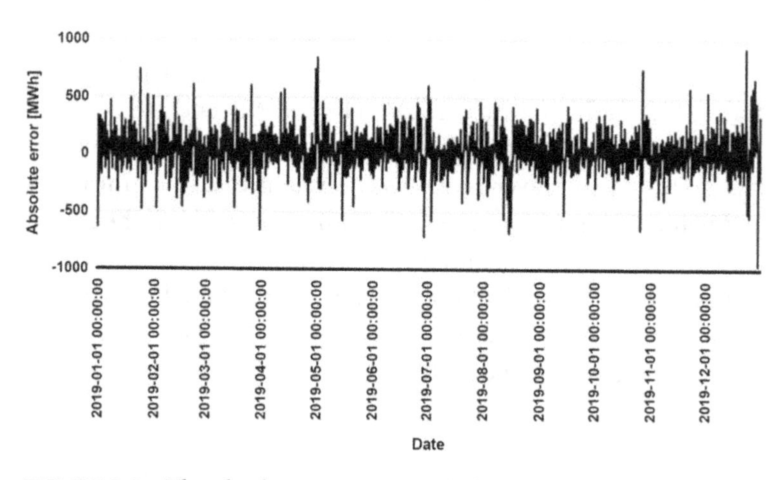

FIGURE 3.9 The absolute error in MWh for the architecture 14

The interpretation of these figures is as follows. The absolute error is represented by the difference between the actual and predicted value of the load. On the other hand, in Figure 3.8, one can see that there are 795 samples that have an absolute error between [−25; 0] MWh and 842 samples that have an absolute error between [0; 25] MWh.

In Ref. [27], the same method has been used to carry out the load forecast for the day ahead and, in that case, it performed

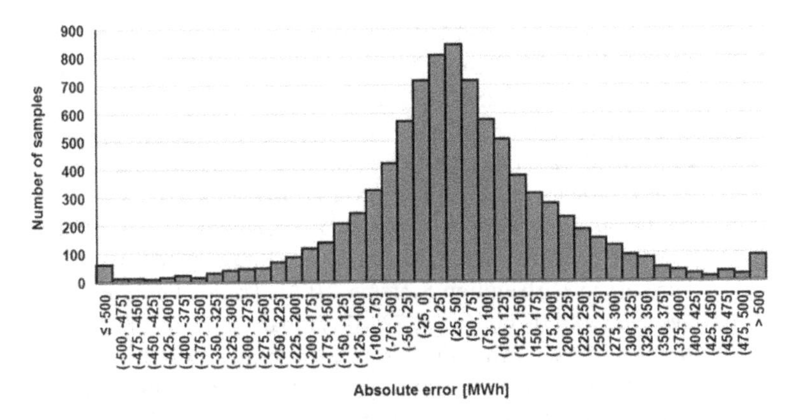

FIGURE 3.10 The error distribution for the architecture 14

better than the instrument used by the transmission operator, even if the CNN was not trained with Big Data. So, one can conclude that the potential of CNNs as far as the load forecast is concerned, is immense.

REFERENCES

[1] Statista – The Statistics Portal for Market Data, Market Research and Market Studies, “Big Data Market Size Revenue Forecast Worldwide from 2011 to 2027,” 2021. [Online]. Available: https://www.statista.com/statistics/254266/global-big-data-market-forecast/

[2] I. Ilin, A. Klimin and A. Shaban, “Features of Big Data Approach and New Opportunities of BI-Systems in Marketing Activities,” E3S Web of Conferences, vol. 110, St. Petersburg, Russia, 2019, pp. 1–9.

[3] Y. Wang, Q. Chen, T. Hong and C. Kang, “Review of Smart Meter Data Analytics: Applications, Methodologies, and Challenges,” *IEEE Transactions on Smart Grid*, vol. 10, no. 3, pp. 3125–3148, May 2019.

[4] J. Xie, T. Hong and J. Stroud, “Long-Term Retail Energy Forecasting with Consideration of Residential Customer Attrition,” *IEEE Transactions on Smart Grid*, vol. 6, no. 5, pp. 2245–2252, September 2015.

[5] T. Hong, P. Wang and L. White, "Weather Station Selection for Electric Load Forecasting," *International Journal of Forecasting*, vol. 31, no. 2, pp. 286–295, 2015.

[6] S. Fan and R. J. Hyndman, "Short-Term Load Forecasting Based on a Semi-Parametric Additive Model," *IEEE Transactions on Power System*, vol. 27, no. 1, pp. 134–141, February 2012.

[7] N. Ding, C. Benoit, G. Foggia, Y. Bésanger and F. Wurtz, "Neural Network-Based Model Design for Short-Term Load Forecast in Distribution Systems," *IEEE Transactions on Power Systems*, vol. 31, no. 1, pp. 72–81, January 2016.

[8] X. Sun et al., "An Efficient Approach to Short-Term Load Forecasting at the Distribution Level," *IEEE Transactions on Power Systems*, vol. 31, no. 4, pp. 2526–2537, July 2016.

[9] R. E. Edwards, J. New and L. E. Parker, "Predicting Future Hourly Residential Electrical Consumption: A Machine Learning Case Study," *Energy and Buildings*, vol. 49, pp. 591–603, June 2012.

[10] H. Chitsaz, H. Shaker, H. Zareipour, D. Wood and N. Amjady, "Short-Term Electricity Load Forecasting of Buildings in Microgrids," *Energy and Buildings*, vol. 99, pp. 50–60, July 2015.

[11] A. Tascikaraoglu and B. M. Sanandaji, "Short-Term Residential Electric Load Forecasting: A Compressive Spatio-Temporal Approach," *Energy and Buildings*, vol. 111, pp. 380–392, January 2016.

[12] C.-N. Yu, P. Mirowski and T. K. Ho, "A Sparse Coding Approach to Household Electricity Demand Forecasting in Smart Grids," *IEEE Transactions on Smart Grid*, vol. 8, no. 2, pp. 738–748, March 2017.

[13] P. Li, B. Zhang, Y. Weng and R. Rajagopal, "A Sparse Linear Model and Significance Test for Individual Consumption Prediction," *IEEE Transactions on Power Systems*, vol. 32, no. 6, pp. 4489–4500, November 2017.

[14] M. Chaouch, "Clustering-Based Improvement of Nonparametric Functional Time Series Forecasting: Application to Intra-Day Household-Level Load Curves," *IEEE Transactions on Smart Grid*, vol. 5, no. 1, pp. 411–419, Jan. 2014.

[15] Y.-H. Hsiao, "Household Electricity Demand Forecast based on Context Information and User Daily Schedule Analysis from Meter Data," *IEEE Transactions on Industrial Informatics*, vol. 11, no. 1, pp. 33–43, February 2015.

[16] P. Wang, B. Liu and T. Hong, "Electric Load Forecasting with Recency Effect: A Big Data Approach," *International Journal of Forecasting*, vol. 32, no. 3, pp. 585–597, 2016.

[17] J. Xie, Y. Chen, T. Hong and T. D. Laing, "Relative Humidity for Load Forecasting Models," *IEEE Transaction on Smart Grid*, vol. 9, no. 1, pp. 191–198, January 2018.

[18] J. Xie and T. Hong, "Wind Speed for Load Forecasting Models," *Sustainability*, vol. 9, no. 5, pp. 795, 2017.

[19] T. Hong, P. Pinson and S. Fan, "Global Energy Forecasting Competition 2012," *International Journal of Forecasting*, vol. 30, no. 2, pp. 357–363, 2014.

[20] J. J. M. Moreno, A. P. Pol, A. S. Abad and B. C. Blasco, "Using the R-MAPE Index as a Resistant Measure of Forecast Accuracy," *Psicothema*, vol. 25, no. 4, pp. 500–506, 2013.

[21] S. Kim and H. Kim, "A New Metric of Absolute Percentage Error for Intermittent Demand Forecasts," *International Journal of Forecasting*, vol. 32, no. 3, pp. 669–679, 2016.

[22] T. Hong, J. Wilson and J. Xie, "Long Term Probabilistic Load Forecasting and Normalization with Hourly Information," *IEEE Transactions on Smart Grid*, vol. 5, no. 1, pp. 456–462, Jan. 2014.

[23] "PJM load forecast report January 2015 prepared by PJM resource adequacy planning department", 2015, [online] Available: https://www.pjm.com/-/media/library/reportsnotices/load-forecast/2015-load-forecast-report.ashx?la=en

[24] D. Gan, Y. Wang, S. Yang and C. Kang, "Embedding Based Quantile Regression Neural Network for Probabilistic Load Forecasting," *Journal of Modern Power Systems and Clean Energy*, vol. 6, no. 2, pp. 244–254, 2018.

[25] J. Black, A. Hoffman, T. Hong, J. Roberts and P. Wang, "Weather Data for Energy Analytics: From Modeling Outages and Reliability Indices to Simulating Distributed Photovoltaic Fleets," *IEEE Power Energy Mag.*, vol. 16, no. 3, pp. 43–53, May/Jun. 2018.

[26] H. H. Aghdam and E. J. Heravi, *Guide to Convolutional Neural Networks*, New York, NY, USA: Springer, pp. 85–127, 2017.

[27] A. M. Tudose, D. O. Sidea, I. I. Picioroaga, V. A. Boicea and C. Bulac, "A CNN Based Model for Short-Term Load Forecasting: A Real Case Study on the Romanian Power System," 2020 55th

International Universities Power Engineering Conference (UPEC), Turin, Italy, 2020, pp. 1–6.

[28] D. P. Kingma and J. L. Ba, "ADAM: A Method for Stochastic Optimization," 3rd International Conference for Learning Representations, San Diego, CA, USA, 2015, pp. 1–15.

CHAPTER 4

Conclusions and Further Research Directions

As already stated in Chapter 1, the load forecast is closely related to the concept of "Industry 4.0." That is, essentially, because an accurate power load forecast brings efficiency in energy consumption and generation as well as important economic benefits. At the same time, Big Data is the concept that unites both "Industry 4.0" and the load forecast. Based on the discovered correlations between the currently available data input, new knowledge can be generated. Some have even argued that the correlation analyses will force the classical causality studies to become practically obsolete [1]. Seen from another perspective, Big Data can very much improve the availability of power plants. And, not only this; Big Data can also help in developing new lines of business, even when it comes to the power sector. This is also carried out through:

- improved and fast market research;
- reduced maintenance costs and working hours, diminishing thus the fatigue of employees; and
- dynamic pricing.

At this point, one can conclude that in spite of the fairly high number of works relating to various techniques of load forecasting, its accuracy is still low because always the prediction is made using small datasets [2].

As a rule, one can distinguish between the general and the specific factors that influence the accuracy of the load forecast. In the first category, there are: customer behavior, time, the economic aspects, the weather, and randomness. In the second, one has those aspects that are directly related to the adopted method (in this case, related to the CNN accuracy).

Customer behavior is important to be taken into consideration because a first observation leads us to conclude that the clients in the same class have almost identical consumption patterns. The differences between these patterns are not important and, that is why, another idea that can help to improve the forecast accuracy is clusterization, as already mentioned.

The time of the day or the season influence greatly the prediction. For instance, one knows already that during the early working hours of the day, the consumption is not very high [2]. The same happens when the temperature is not very low or very high. Also related to time are the transitions between working and free days (or vice versa). These aspects can be efficiently treated using various types of ANNs, as proposed in this work.

The economy is represented by the prices of fossil fuels that can greatly influence power generation and thus consumption. An accurate prediction model takes into account this price as well.

Finally, randomness is related to events that are not characteristic of a certain area of the grid. So, especially when the load forecast is made at the national level, as in this case, important sport or social events or industrial experiments [2] can alter the regular consumption pattern. This kind of aspect cannot be dealt with in a deterministic way. The only solution in this situation is the statistics.

As far as the specific factors are concerned, in our particular case, the major drawback is that, in the absence of random factors, there is no rule of thumb based on which, the number of the convolutional layers, of the pooling layers as well as the number of filters and their dimension, can be determined. Future research should investigate this more closely because this means implicitly less time. Another important future research direction is to find a way to include simultaneously more than one exogenous variable that influences the prediction accuracy. Usually, the only external factor considered is the temperature. But, obviously, that is not enough. Why is that?

Because if one wants to make a load forecast only for wind, or only for PV, other exogenous variables are useful, like the wind speed variation over the year or the solar irradiance. The same happens in the case of the electricity generated through fossil fuels. In this case, the prices of oil or natural gas over the year play a key role.

A third research direction could be the development of special processing Big Data algorithms, dedicated to power engineering. Anyway, nowadays, the deployment of smart meters, on a large scale, will facilitate the use of Big Data for the ANNs training.

But, in spite of these drawbacks, the load forecast based on CNNs shows very promising results even when not trained with large volumes of datasets.

That is why, one can say, that this particular area of the load forecast has a huge potential and very much can still be done to make it more accurate.

REFERENCES

[1] V. M. Schönberger and K. Cukier, *Lernen mit Big Data: die Zukunft der Bildung*. Munich, Germany: Redline Verlag, 2014.

[2] A. Almalaq and J. J. Zhang, "Deep Learning Application: Load Forecasting in Big Data of Smart Grids," in: *Deep Learning: Algorithms and Applications. Studies in Computational Intelligence, vol 865*, W. Pedrycz and S. M. Chen, Eds., Switzerland: Springer, 2020, pp. 103–128.

Index